THE ULTIMATE

Fluid Pouring & Painting

PROJECT BOOK

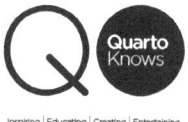

Brimming with creative inspiration, how-to projects, and useful information to enrich your everyday life, Quarto Knows is a favorite destination for those pursuing their interests and passions. Visit our site and dig deeper with our books into your area of interest: Quarto Creates, Quarto Cooks, Quarto Homes, Quarto Lives, Quarto Drives, Quarto Explores, Quarto Gifts, or Quarto Kids.

First Published in 2020 by Quarry Books, an imprint of The Quarto Group,
100 Cummings Center, Suite 265-D, Beverly, MA 01915, USA.
T (978) 282-9590 F (978) 283-2742 QuartoKnows.com

Quarry Books titles are also available at discount for retail, wholesale, promotional, and bulk purchase. For details, contact the Special Sales Manager by email at specialsales@quarto.com or by mail at The Quarto Group, Attn: Special Sales Manager, 100 Cummings Center, Suite 265-D, Beverly, MA 01915, USA.

ISBN: 978-1-63159-763-3

Digital edition published in 2020
eISBN: 978-1-63159-764-0

Library of Congress Cataloging-in-Publication is available

Design: Samantha J. Bednarek
Page Layout: Samantha J. Bednarek

THE ULTIMATE

Fluid Pouring & Painting

PROJECT BOOK

Inspiration and Techniques for Using Alcohol Inks, Acrylics, Resin, and More

Create Colorful Paintings, Resin Coasters,
Agate Slices, Vases, Vessels & More

Jane Monteith

QUARRY

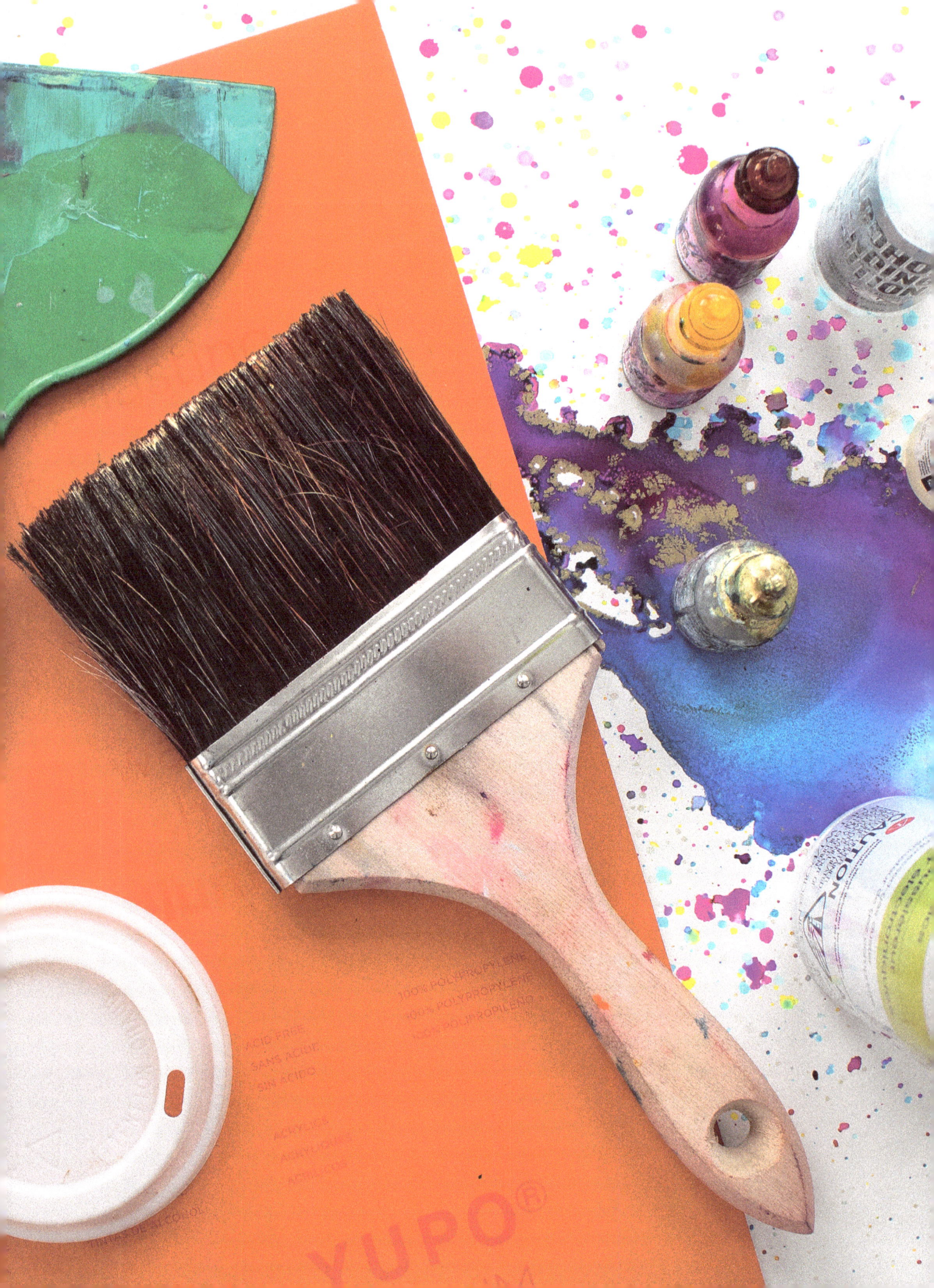

CONTENTS

AN INVITATION TO CREATE

I have been in love with art, creativity, and color for as long as I can remember. I immigrated from England to Canada when I was ten years old. Growing up, I incorporated creativity into all aspects of my life, but I didn't truly begin exploring art as a paying career until my mid-thirties. I always believed an office job would be where I would stay.

Yet something or someone was always trying to pull me into a creative path and when a small window of opportunity opened, I decided to take a leap of faith and pursue painting as a means of making a living. I started my own business and traveled around my province of Ontario, painting mascots and logos on vinyl crash mats in public and high school gyms with highly pigmented screening inks. I did this for more than a decade until I grew tired of the traveling and wished for a change.

As a self-taught artist, I began immersing myself in all forms of art, exploring new art mediums and experimenting with new techniques. It was at this time that I first stumbled across alcohol inks. They were the first fluid medium that I fell for. I loved the vibrant color and saturation of alcohol inks, because they reminded me of the bold colors of screening inks. It was then that I knew I had found my true love and I decided to passionately pursue everything fluid art had to offer. Since then, I've shared my love of fluid art with thousands through social media and my online art courses. This book walks you through some of my favorite fluid art projects, which I know will become your favorites too.

WE WERE BORN TO BE CREATIVE!

I believe that all of us are creative. Some of us just haven't found the key to unlocking what creativity can offer. I believe it to be an important part of life, just like good nutrition and exercise are essential for overall well-being. When people incorporate creativity into their lives, good things happen. You begin to use another side of your brain that assists in other aspects of your daily routine. Creative activities can help reduce stress by putting your mind at ease, allowing you to think analytically (and visually) and make better decisions. Creativity can take us to a state of joy, help us express emotions, build confidence, and release hormones that make us happy. Exercising your sense of creativity can take you away to another world, where you leave behind the daily grind of work, negativity, and pressure. It is something everyone should explore. Fluid art is just one way you can begin exercising your creative muscle, all while creating some beautiful art along the way! So, carve out some time for yourself, won't you? You won't regret it. I can't wait for you to get started.

—Jane

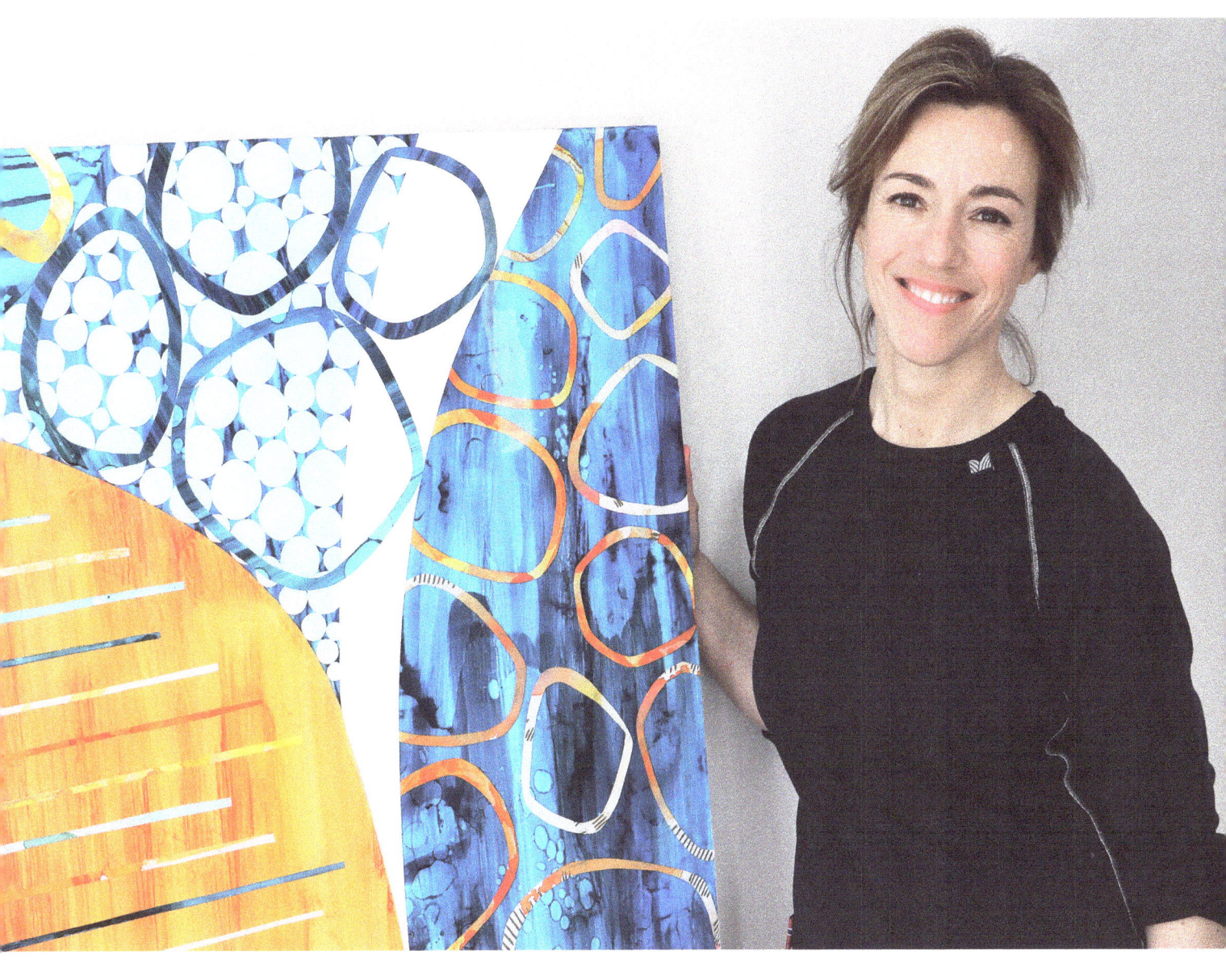

YOU CAN'T USE UP CREATIVITY. THE MORE YOU USE, THE MORE YOU HAVE. —*Maya Angelou*

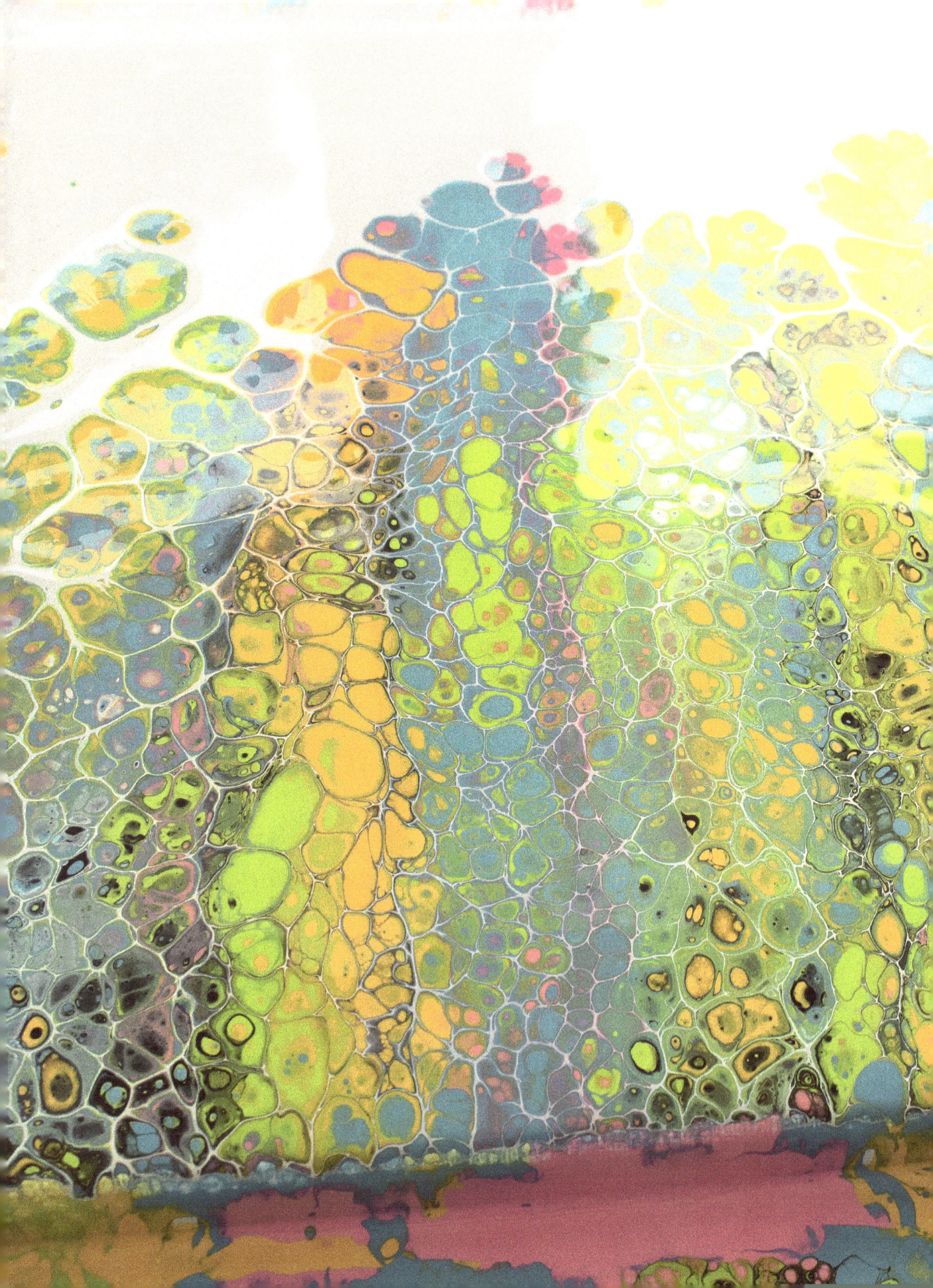

INTRODUCTION

You've picked up this book, perhaps out of curiosity, the desire to learn a new art medium or new techniques, or just for some plain old fun. I'm so excited that you did! The projects and tips you'll find on the following pages will satisfy all those things. This book is the ultimate guide dedicated to the art of fluid pouring and painting. It is an art form that has exploded over the past several years for reasons you'll come to understand as you explore the pages of this book. And because, well, it's so addictive! Many people fall in love with the colors, vibrancy, and random beauty that can be achieved with fluid art.

Different products and tools can be used to create beautiful paintings and décor in a variety of styles. In this book you will find inspiring photos, step-by-step project techniques and instruction, and tips to help you create satisfying art that you will want to keep and display or give to your family and friends. These projects are ideal for beginners or longtime practicing artists. No prerequisites are needed. You don't need an art degree, tons of experience creating art, or knowledge about the art world in order to begin. All you need is some patience and the willingness to learn. You'll be having too much fun to even realize you're learning a new art form.

BEFORE WE BEGIN...

Before diving into the projects, I would like to share a few things about what you can expect to find in this book. Aside from pure inspiration, I share many resources, tips, tricks, and suggestions.

While creating is fun, there can sometimes be an element of frustration that can leave you wondering what you did wrong. I often get requests through my Instagram account from people who want to fix an issue or know why it happened in the first place. There are many common problems that can occur, so I have included a resin troubleshooting section to help you prevent or remedy these issues.

After testing and experimenting with a lot of products, I have come to rely on specific brands that I trust to consistently provide great results. These brands may or may not be accessible to you, depending on where you live. Fortunately, there are numerous other products on the market that do the same thing and I do my best to list alternative materials for each project. To ensure you find exactly what you need, I've included a resource list at the back of the book.

Lastly, I would like to point out that some of the materials and products used in this book should only be used in a well-ventilated area. Wearing a mask and gloves, especially if you're prone to allergies or are overly sensitive to smells, is a good idea as well.

Are you ready? Then let's get creative!

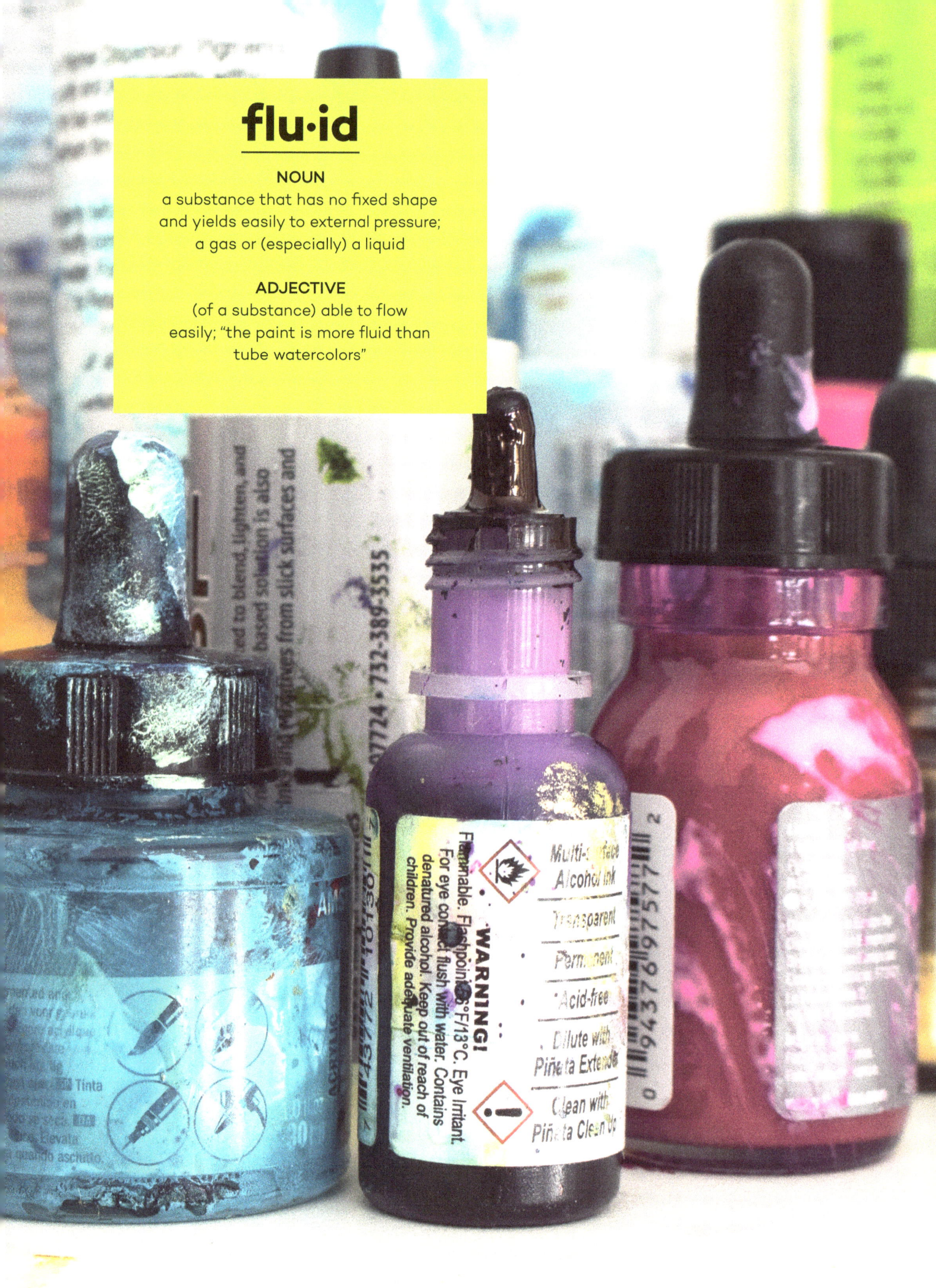

flu·id

NOUN
a substance that has no fixed shape
and yields easily to external pressure;
a gas or (especially) a liquid

ADJECTIVE
(of a substance) able to flow
easily; "the paint is more fluid than
tube watercolors"

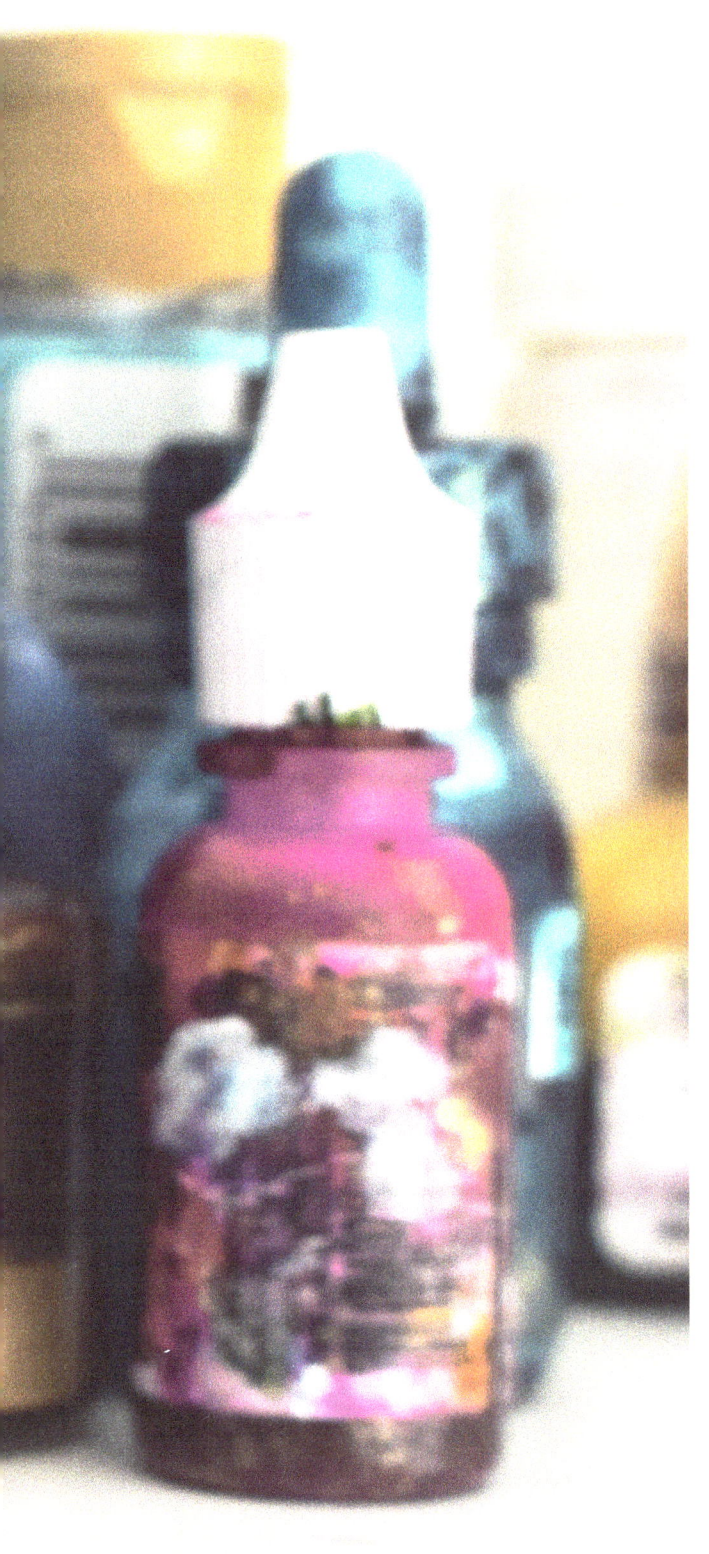

WHAT IS FLUID ART?

Fluid art is a form of art created using an art medium that can be poured, dropped, and manipulated on a substrate without the need for a paintbrush. Using various art tools—including items found around your home—allows you to create different textures and effects, adding a uniqueness and flair to your work. The beauty lies within the free-flowing form the artwork takes from beginning to end. Fluid pouring and painting allows you to relax, enjoy, and embrace the many exciting results that occur as the paint transforms into art.

Have patience. Fluid art can be surprising and immensely satisfying or, when things go wrong, very frustrating. Incorrect color mixing, improper product ratios, or poor techniques can lead to disappointing results. Expect that initially there may be a lot of trial and error. But the great thing about fluid art is that once you get the hang of it, you'll get the desired result and end up with a piece you are happy with. Through experimentation and by following the tips and techniques provided in this book, you will gain confidence, improve your skills, and learn to create beautiful art.

HOW MUCH RESIN?

Determining the right amount of resin mixture for your projects depends on the size of the piece you're working on. When you first start using resin, you may find it difficult to work out just how much to mix. This can result in too little or too much resin. If the brand you're using doesn't offer guidelines within the instructions, a good source to help with calculation is www.artresin. com. This website includes a calculator that allows you to enter the dimensions of your project and receive a recommendation for the required amount in both ounces and milliliters.

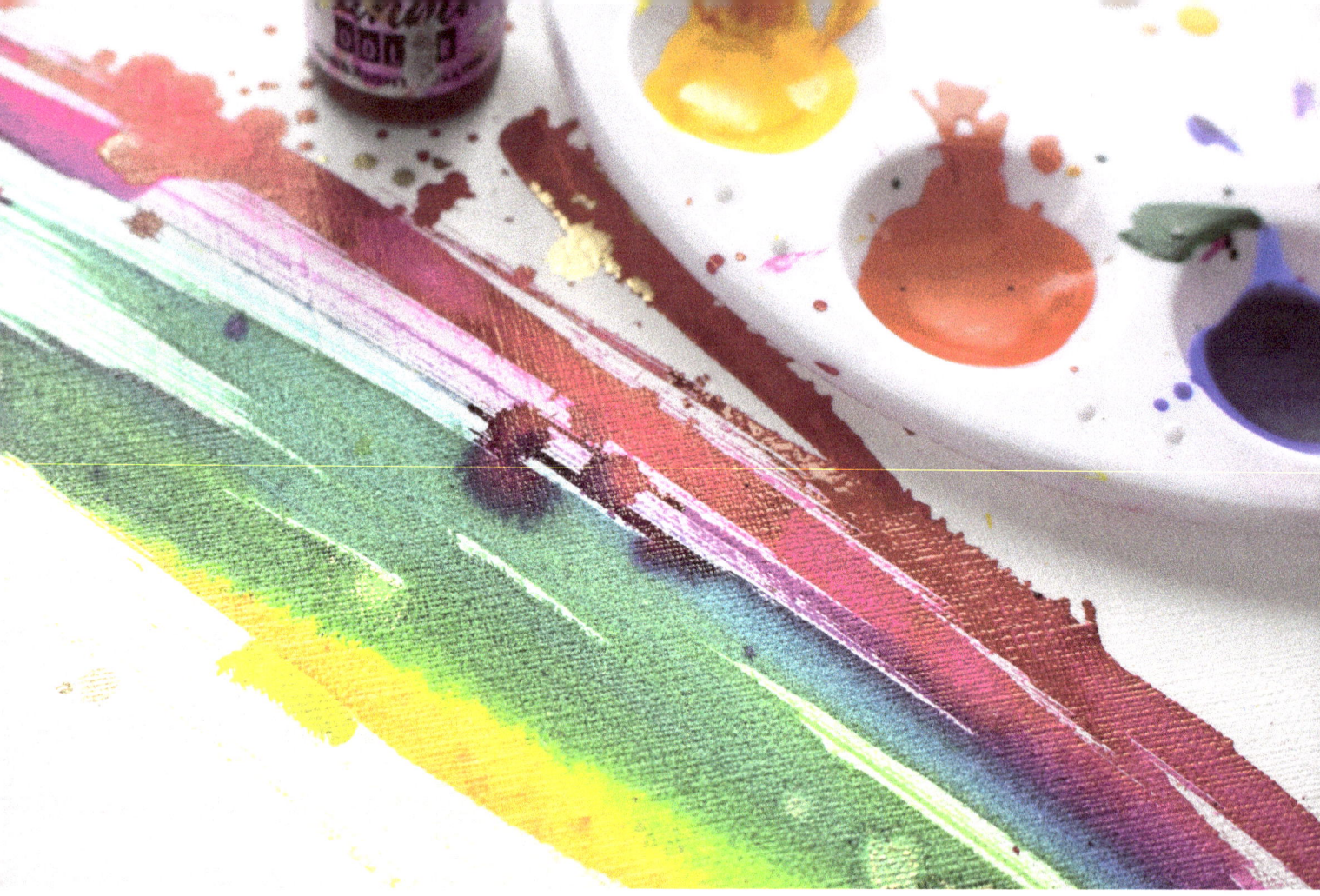

ABOUT COLOR

There is nothing more exciting than diving into an art project when you have all your materials, colors, and tools sitting in front of you. I'm like a kid in a candy store—I just want to open everything up and have at it. But before starting any project, no matter the medium, you must consider color. Color is the most important aspect of art. It's what makes a masterpiece! When you look at a painting you purchased, the home décor item in your living room, or the color on your walls at home, those colors have meaning. Colors can evoke emotions and affect your moods. There are cool colors and warm colors. There are complementary colors and analogous colors. Basically, there are thousands of ways you can combine color to create art. The colors in the projects that follow are fun and vibrant, and they are all thought-out, intentional color palettes.

Choosing colors can be intimidating if you don't understand a bit about color theory. The best way for anyone to gain knowledge of color is to have a color wheel. It is a great way to start experimenting and mixing colors together, and it helps you determine which colors work well together and which ones may result in a muddy mess. My advice for anyone starting out with color is to keep it simple. Start off with two or three colors and take your time.

Once you have experimented with a color wheel, you may still need some color inspiration. I'm always seeking out new combinations that speak to me and Pinterest is a great resource. There are many color palettes that you can pin for inspiration. I have my own color mood board with palettes I created from my own artwork.

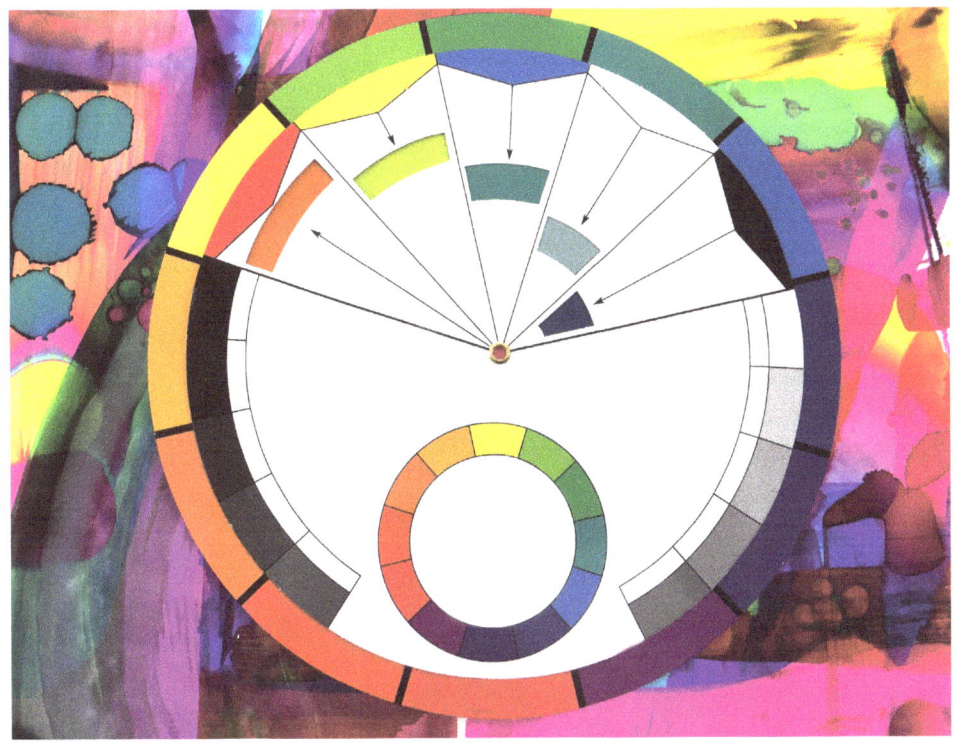

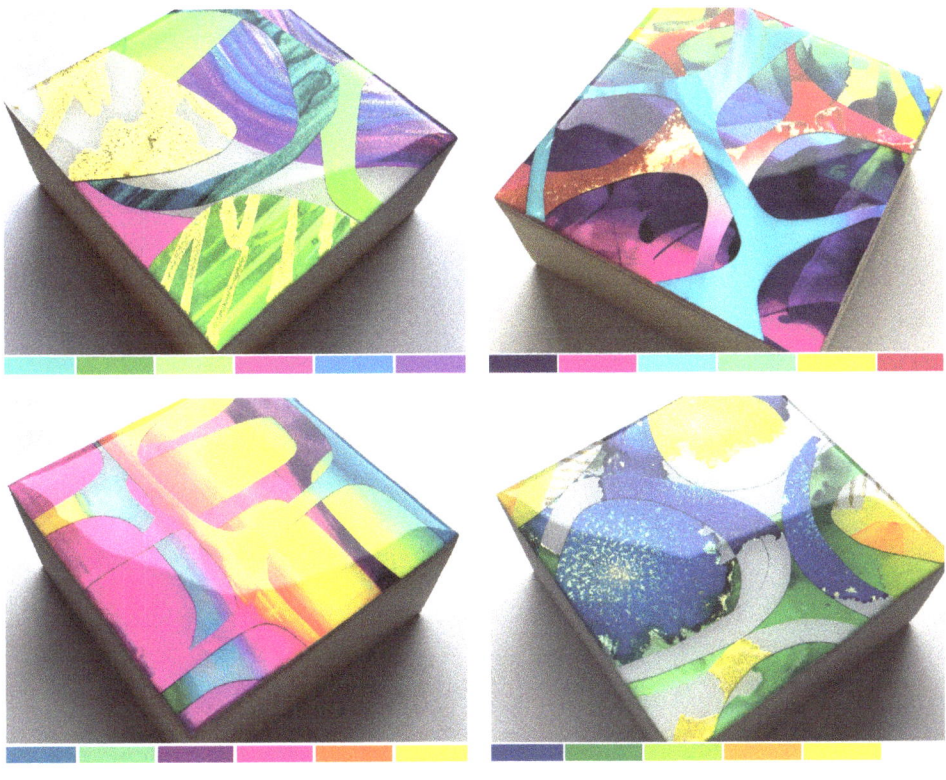

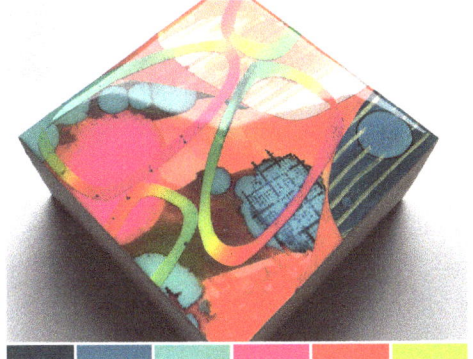

TIPS

- Start with the primary colors (red/blue/yellow) in your tool box so you can create additional colors. Also include white to lighten colors and black to darken them.

- If you create a custom color you love, write it down so you can always look up how you made it.

- To avoid mud, don't mix red and green (green = yellow + blue), as they will create brown.

- Take photos of your color combinations.

- Create color mood boards using Pinterest.

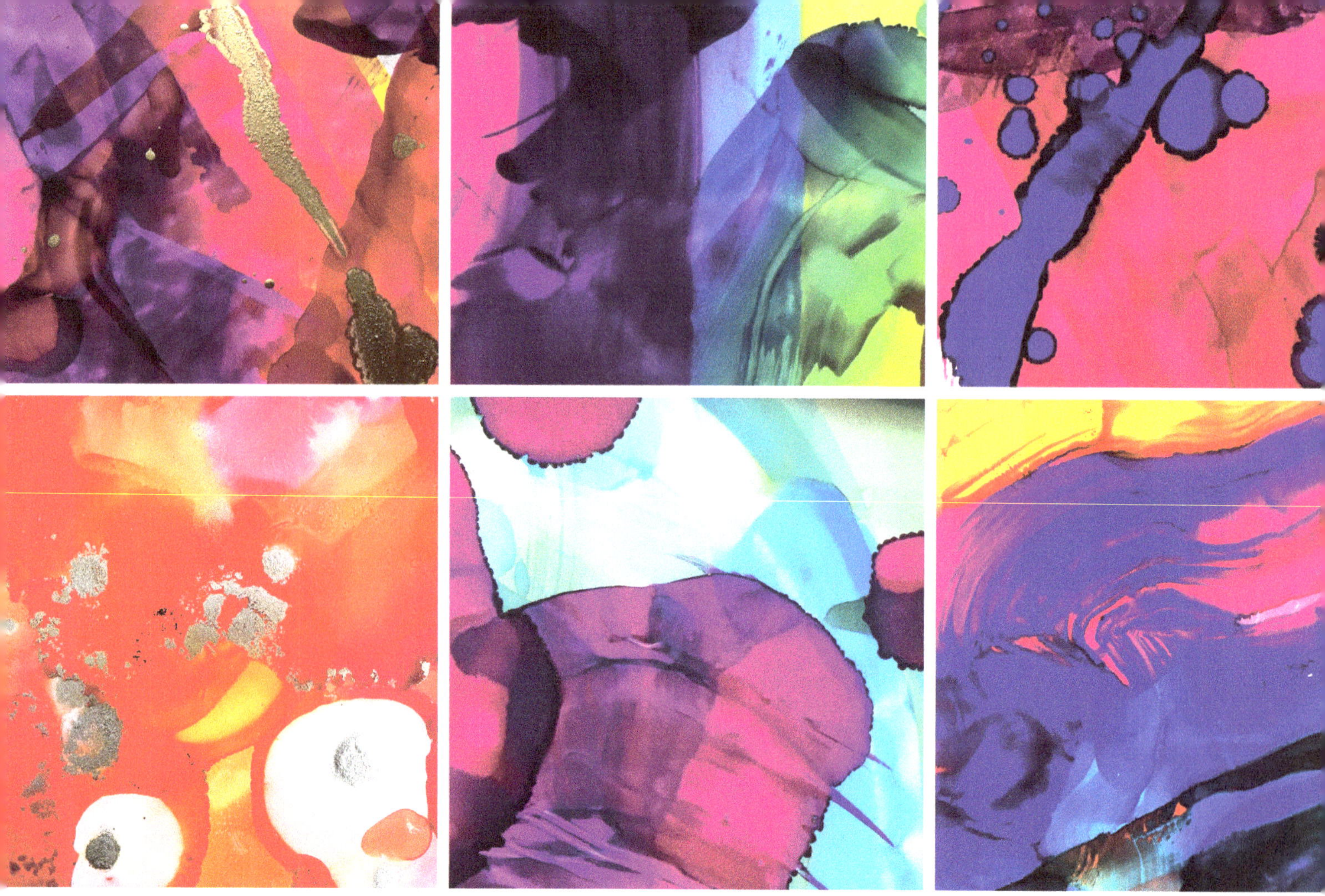

ABOUT ALCOHOL INKS

Alcohol inks are extremely versatile and can be incorporated into many of the projects you'll find throughout this book. They're acid-free, highly saturated dye-based liquids made from colorant suspended in a solution. All dye-based inks are fugitive, which means that if they are subjected to UV exposure over long periods of time, they will ultimately fade. We'll discuss ways to seal and preserve color later on.

Alcohol inks are known for their amazing color and vibrancy. They come in a range of colors, including metallics (gold, silver, copper, and pearl). The metallics are pigment-based inks, which means they are lightfast and will stand the test of time without fading. They add a beautiful rich shine to any painting or poured project. You can also use metallic paint pens to add hand-drawn textures with more control.

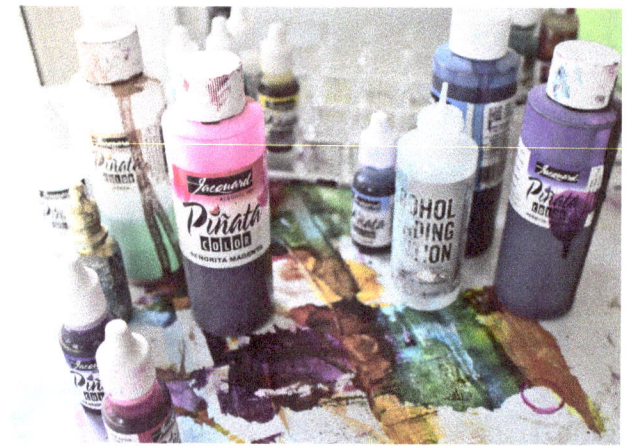

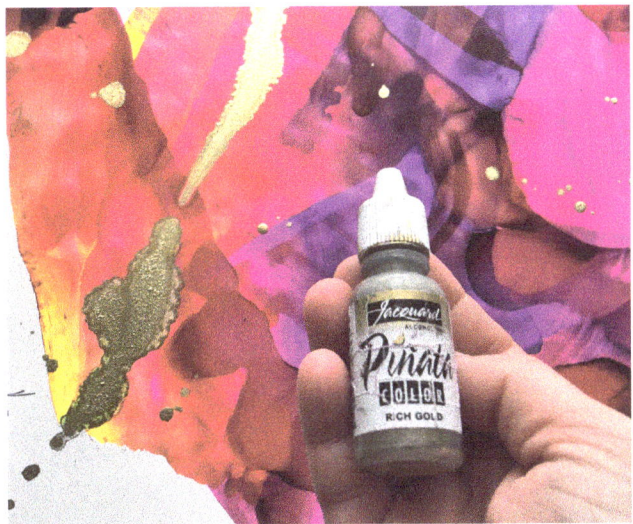
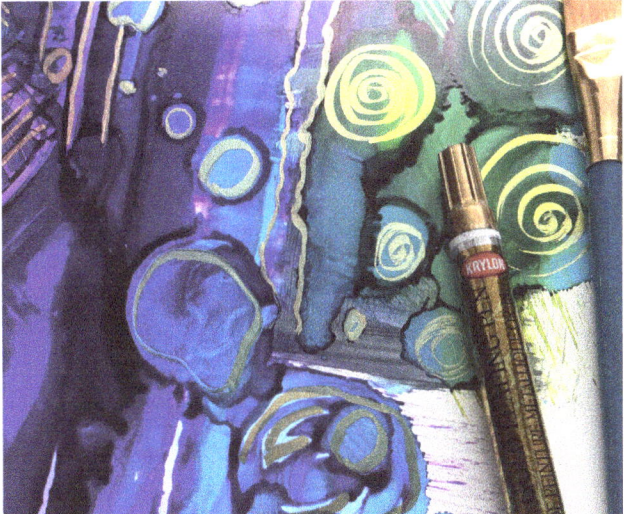

There are many brands of alcohol inks to choose from and experiment with. In addition to the inks, other solutions—such as Tim Holtz Alcohol Blending Solution, Jacquard Claro Extender and/or isopropyl alcohol, or Jacquard Clean Up Solution—will allow you to manipulate and add greater flow, depth, and transparency to your paintings.

BLENDING SOLUTION VS. ISOPROPYL ALCOHOL

Using a blending solution will allow you to extend your working time and to soften and blend your inks. Adding a very small amount of ink to your solution will lighten the intensity of the color. Using isopropyl alcohol of 71 percent or higher will allow you to achieve similar results, but the color will be more of a dry matte with less sheen than you would get from the blending solution. You can buy blending solution from either Tim Holtz (called Blending Solution) or Jacquard (called Claro Extender). If you are sensitive to strong odors, use Jacquard Clean Up Solution instead of alcohol. Both are great for cleaning up brushes and tools as well.

Alcohol inks work best on nonporous surfaces such as glossy papers, glass, metal, tiles, or anything with a slick surface that will cause the ink to sit on top and move around on instead of seeping into the substrate. You can use alcohol inks on a porous substrate such as canvas or wood, but you must prepare the surface so that it is impenetrable.

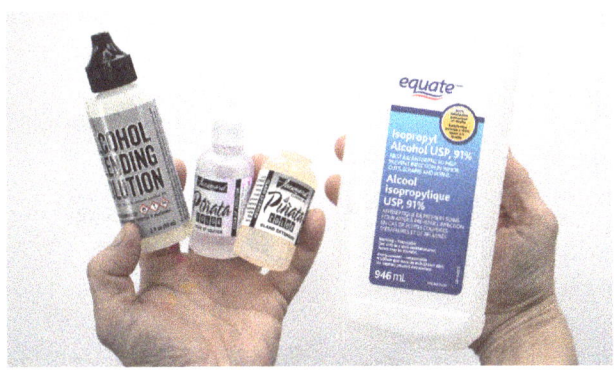

The most popular substrate for alcohol inks is Yupo paper, a plastic composite material made of polypropylene. It looks just like regular paper, but is smooth and does not rip, tear, or deteriorate. It is readily available in art stores and online.

TIP

Can't find blending solution? Make your own by combining approximately 2 ounces (60 ml) of 91% isopropyl alcohol and 2 to 3 drops of pure glycerin.

ALCOHOL INK ABSTRACT PAINTINGS IN THREE STYLES

In this project, you will use different techniques to create several basic styles of abstract paintings on Yupo paper. You'll be surprised at what you can create with alcohol inks in mere minutes. While they can be tamed to paint detailed and realistic paintings, the fun is in the flow. The key with this project is to just have fun and experiment with the ways you can manipulate the inks. By dropping inks onto the paper, adding solutions, or moving them around by picking up the paper or using fun tools, you will change the outcome of the painting each time.

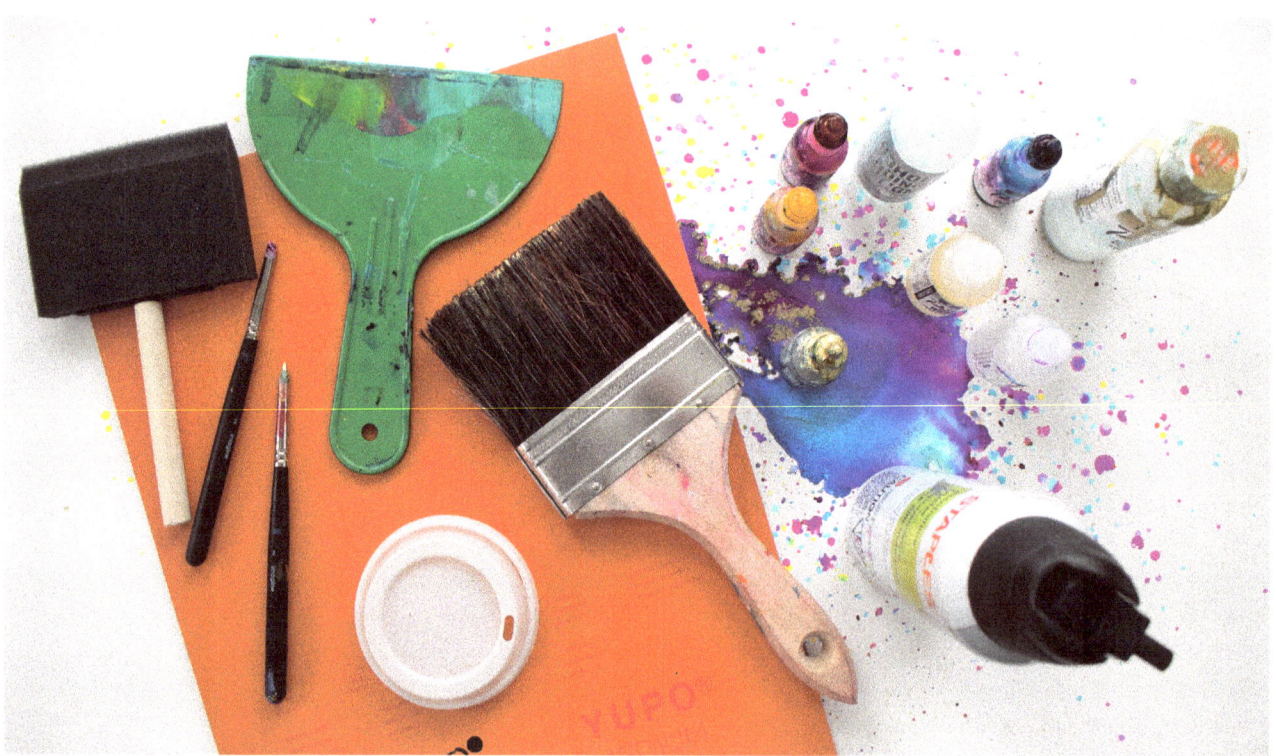

MATERIALS

alcohol inks

Yupo paper

isopropyl alcohol (71% or higher)

blending solution

*Jacquard Clean Up Solution (optional)

Optional: paint-manipulating tools such as old paintbrushes, small round-tip brush, dropper tool, Fantastix, Doodle Stix, plastic scraper, plastic ruler, plastic coffee cup lid, compressed air can, straw, and hair dryer

SAFETY

- Always work in a well-ventilated area.
- Use a mask that is certified against organic vapor.
- Wear powder-free nitrile gloves.
- Do not spray alcohol inks.
- Alcohol inks are flammable, so keep them away from excess heat and open flames.

TIPS

- Use plastic tools so you can easily clean them with isopropyl alcohol and reuse them.
- Isopropyl alcohol of less than 71 percent concentration will create a grainy effect due to its water content.
- If you are sensitive to strong odors, replace isopropyl alcohol with *Jacquard Clean Up Solution.
- Cleanup for hands: mix baking soda with a few drops of dish soap to create a paste when hands are rubbed together.
- Cleanup for brushes: isopropyl alcohol, Jacquard Clean Up Solution
- When using brushes with alcohol inks, don't use the same brush in different colors, even after cleaning.

YUPO SUBSTITUTES

- Glossy photo paper, reverse side
- Vellum
- Clear Dura-Lar film
- Modeling film

AVAILABLE BRANDS

- Various alcohol ink brands: Jacquard Piñata, Tim Holtz Ranger, Brea Reese, Copic Various Ink Refill, Spectrum Noir
- Blending Solution: Jacquard Claro Extender or Tim Holtz Blending Solution

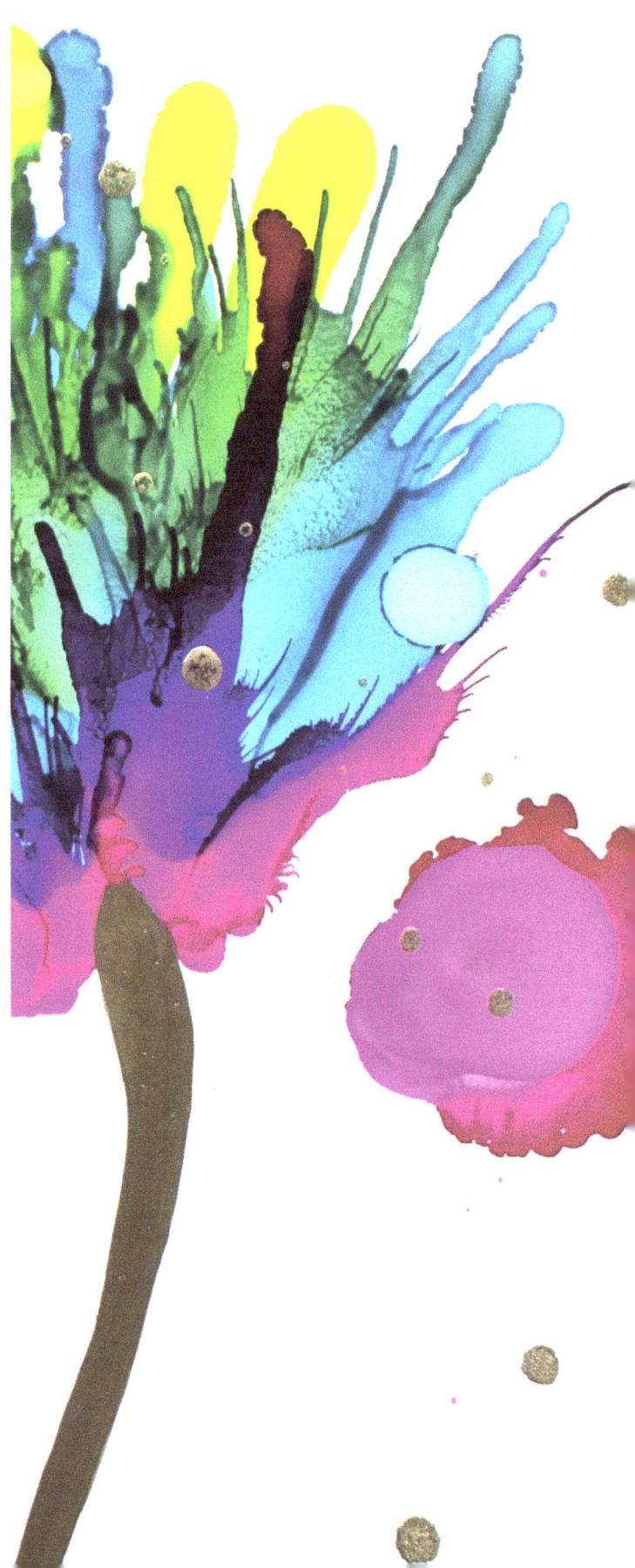

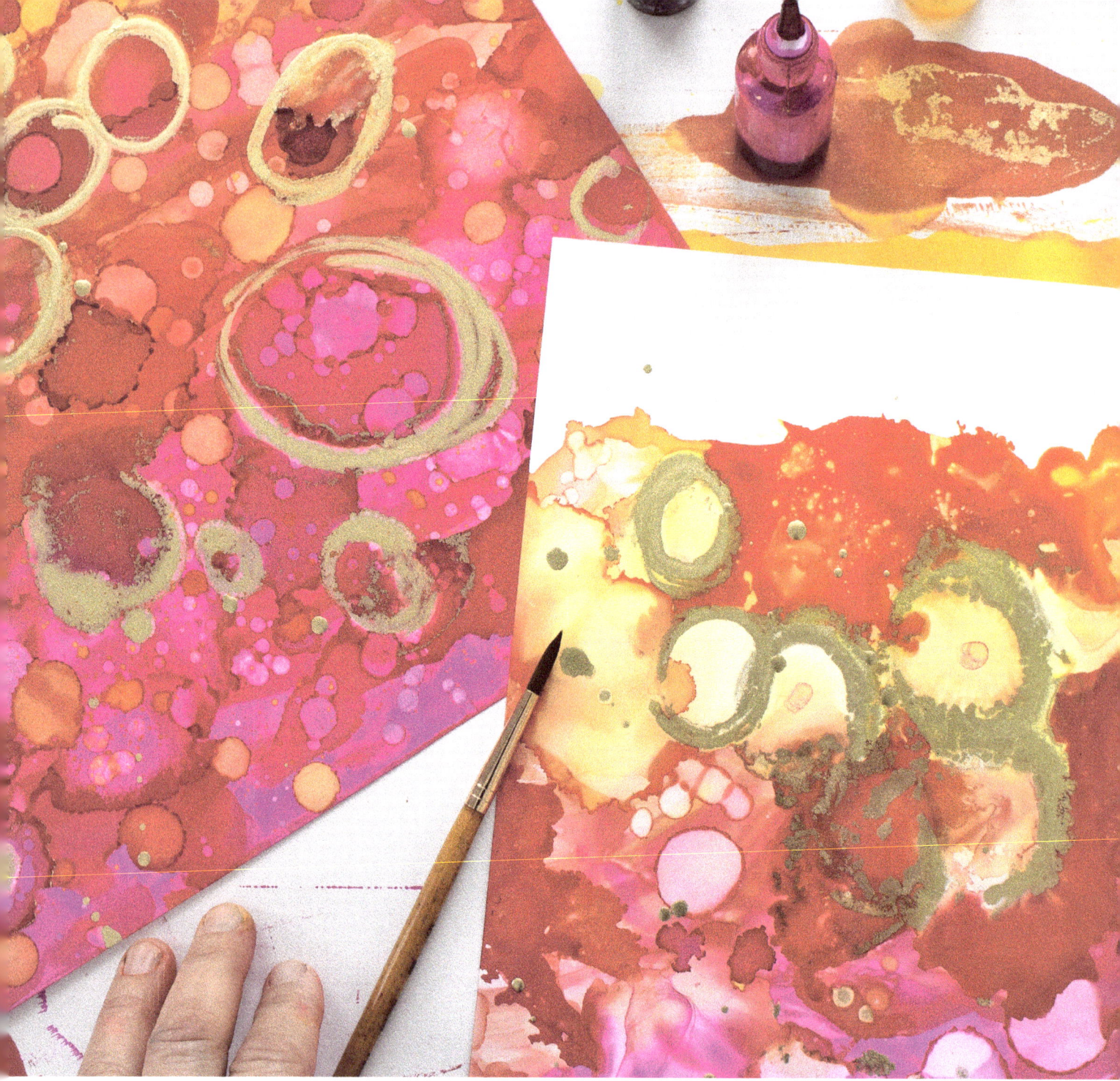

STYLE 1

PINK AND PRETTY

In this first painting you will use only a plastic coffee cup lid as a tool. Pick two or three alcohol ink colors of your choice (colors used in example: Señorita Magenta, Sunbright Yellow, and Rich Gold by Jacquard) and use either blending solution or isopropyl alcohol for flow. Remember to wear gloves!

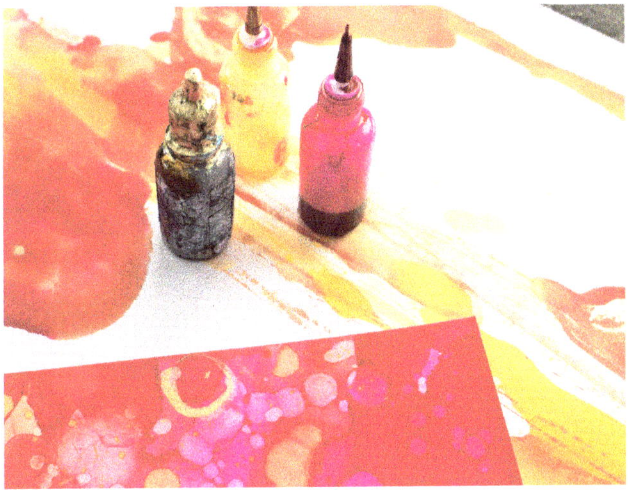

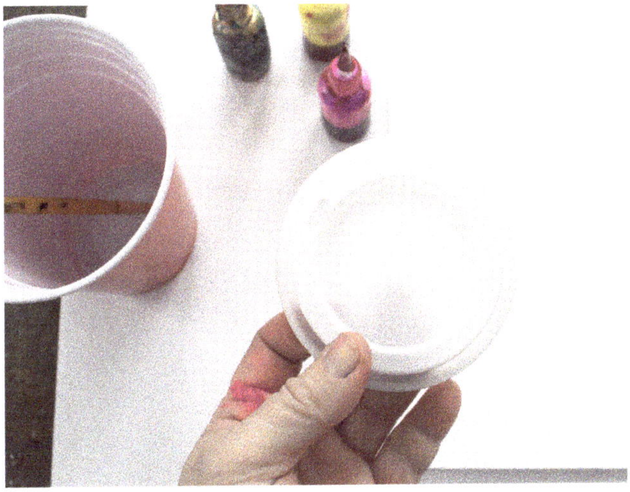

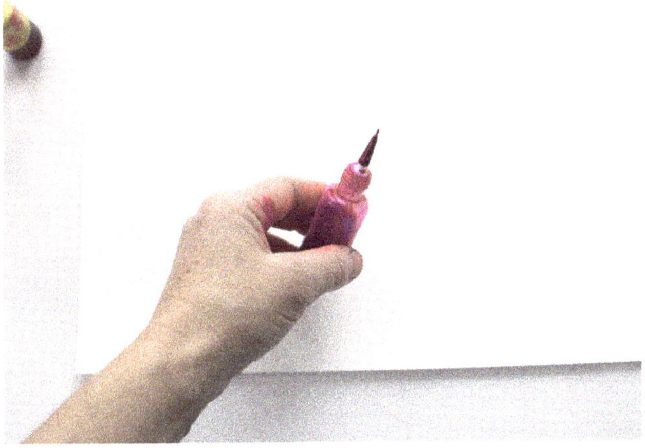

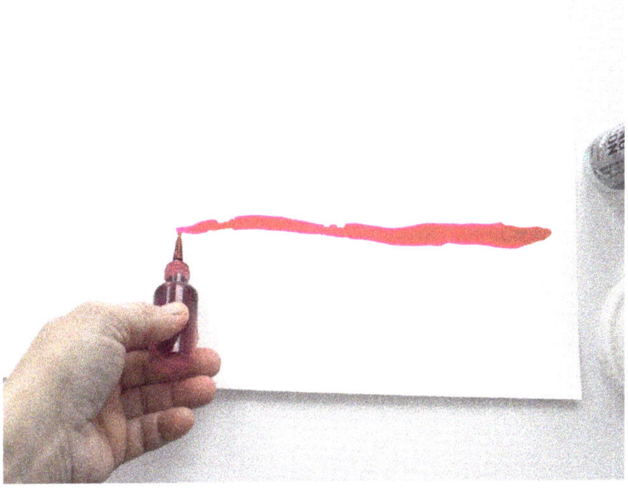

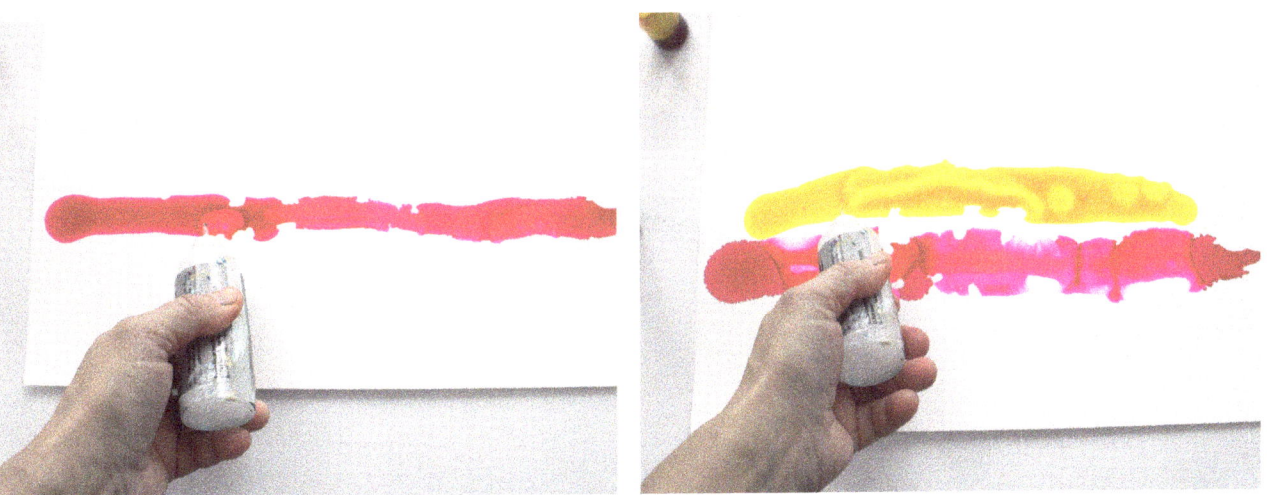

1. Begin with a few drops or light squeeze of ink onto your Yupo paper. Less is more, and you will find a little goes a long way. It's easy to get carried away with alcohol inks and add too much.

2. Add blending solution to help the ink flow across the page. Repeat the process with a second color.

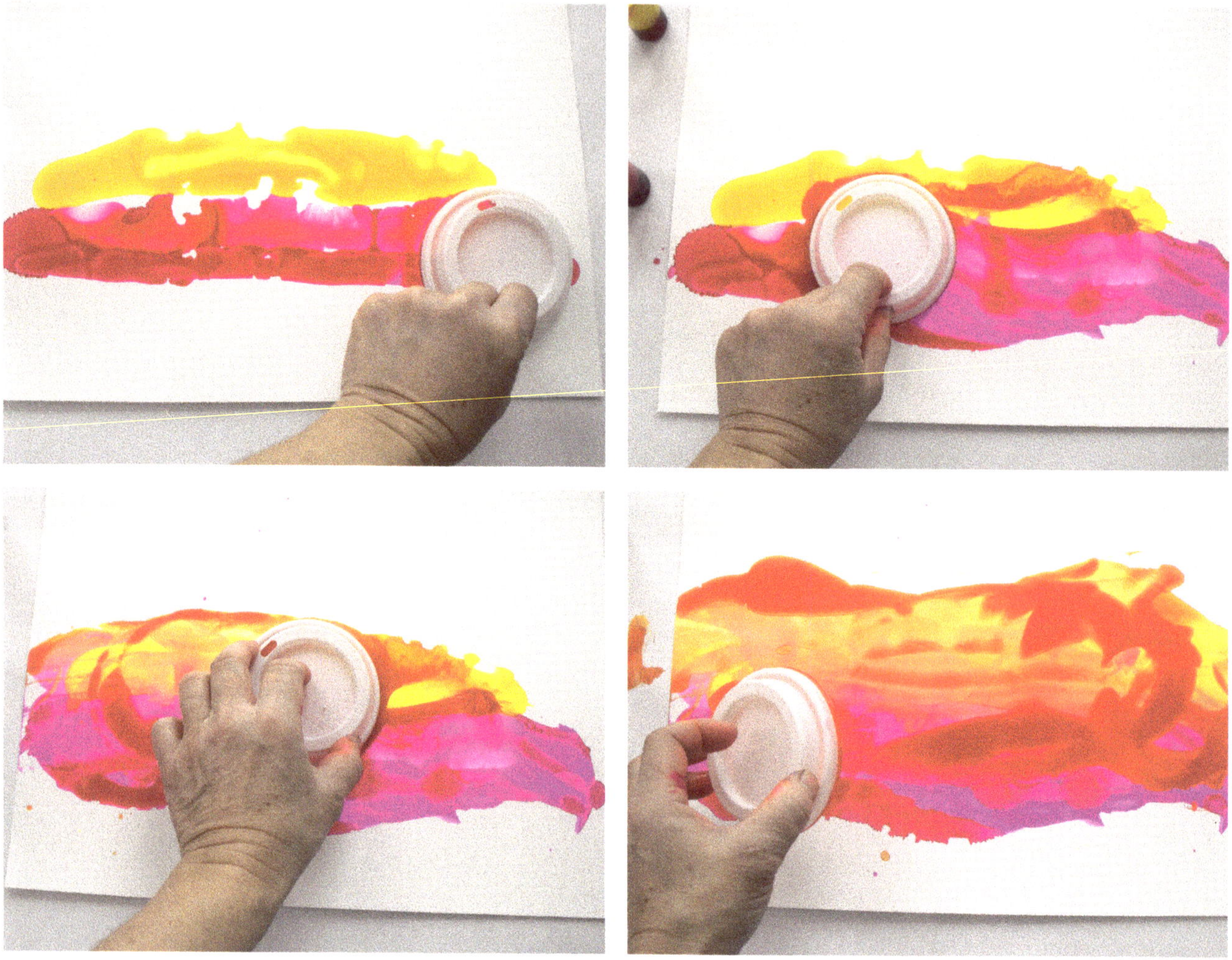

3. Drag and move the inks and solution around with a plastic coffee cup lid.

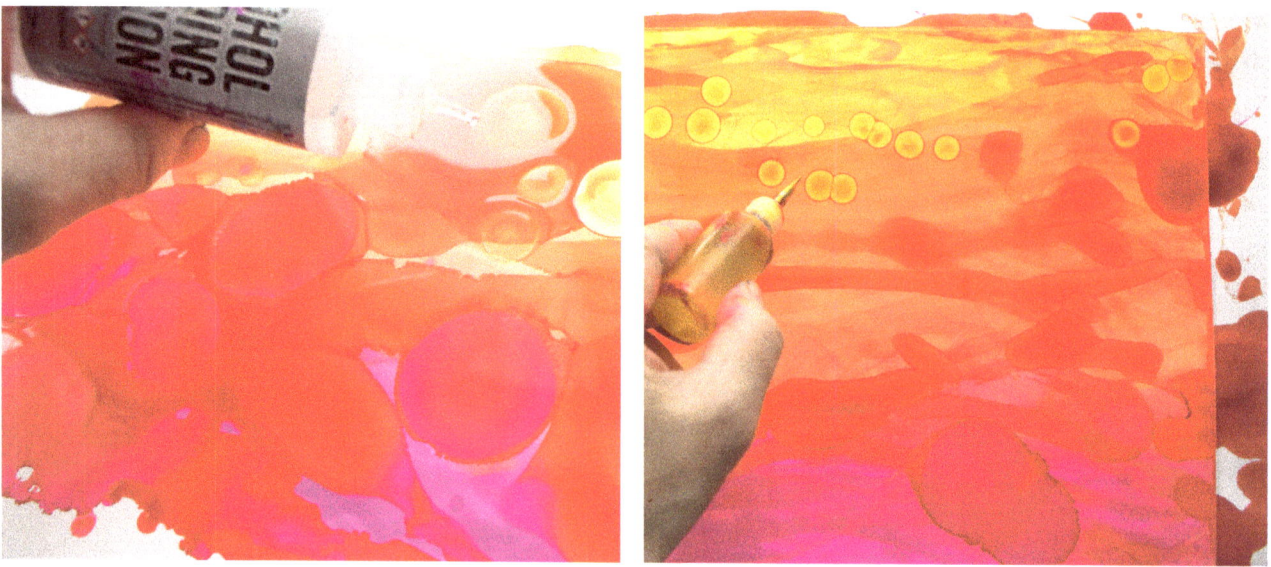

4. Allow the layer to dry for a few minutes before adding more solution and inks. Create texture with each layer by adding a new color or the same color to deepen the shade. Adding a new color on top of a partially dried one will change the color. (Example: Adding yellow onto pink will make orange.)

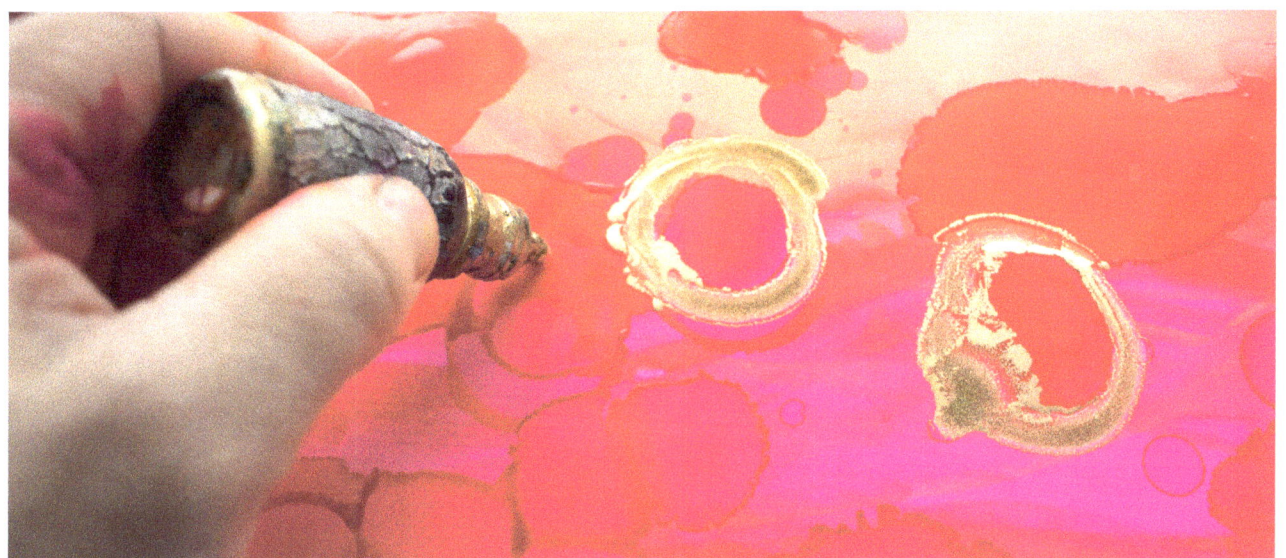

5. Use the nib of the bottle as a tool to move or draw inks onto the Yupo. Shake drops or use a dropper tool to add another element to your painting. When using metallic inks, shake the bottles well before dropping them onto your art.

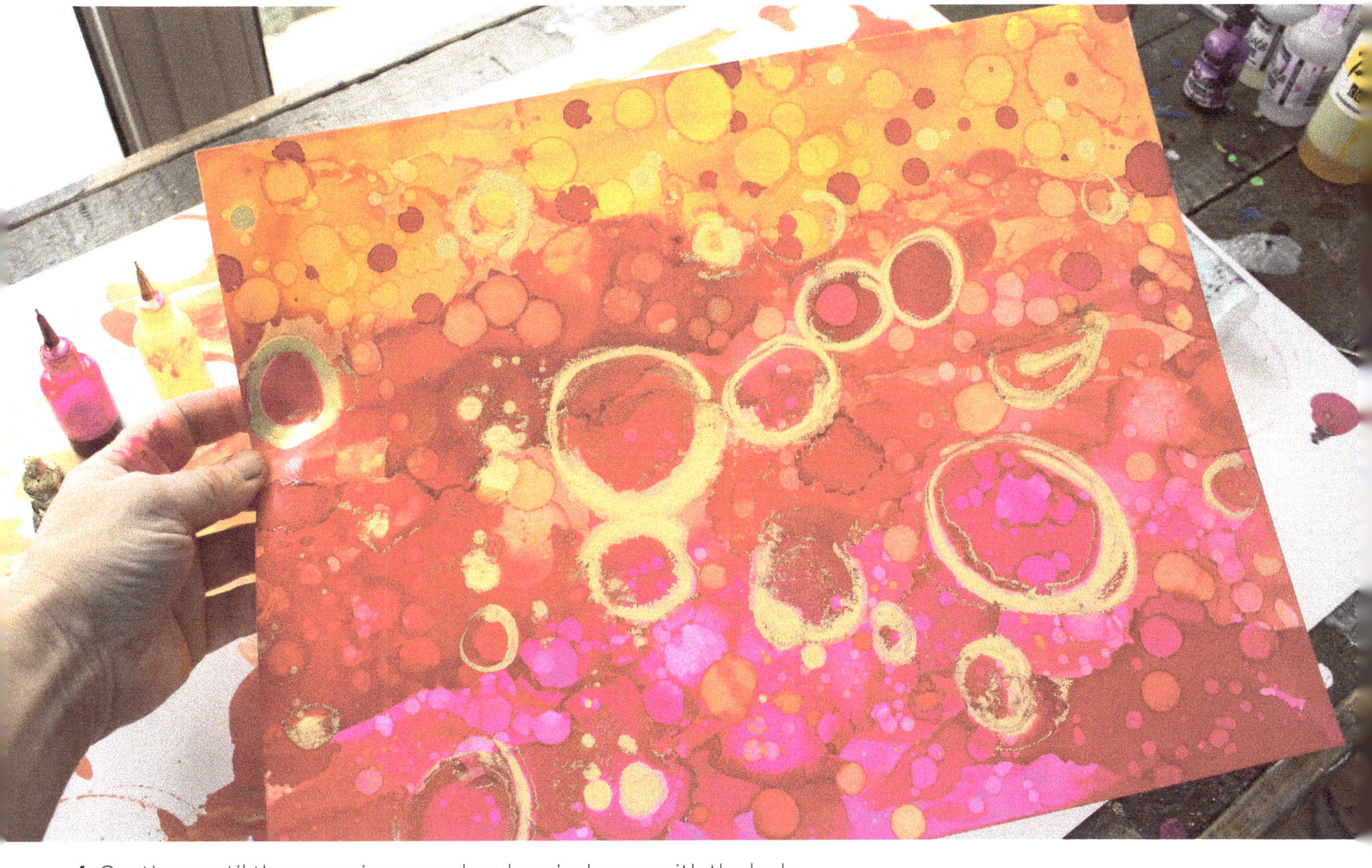

6. Continue until the paper is covered and you're happy with the look.

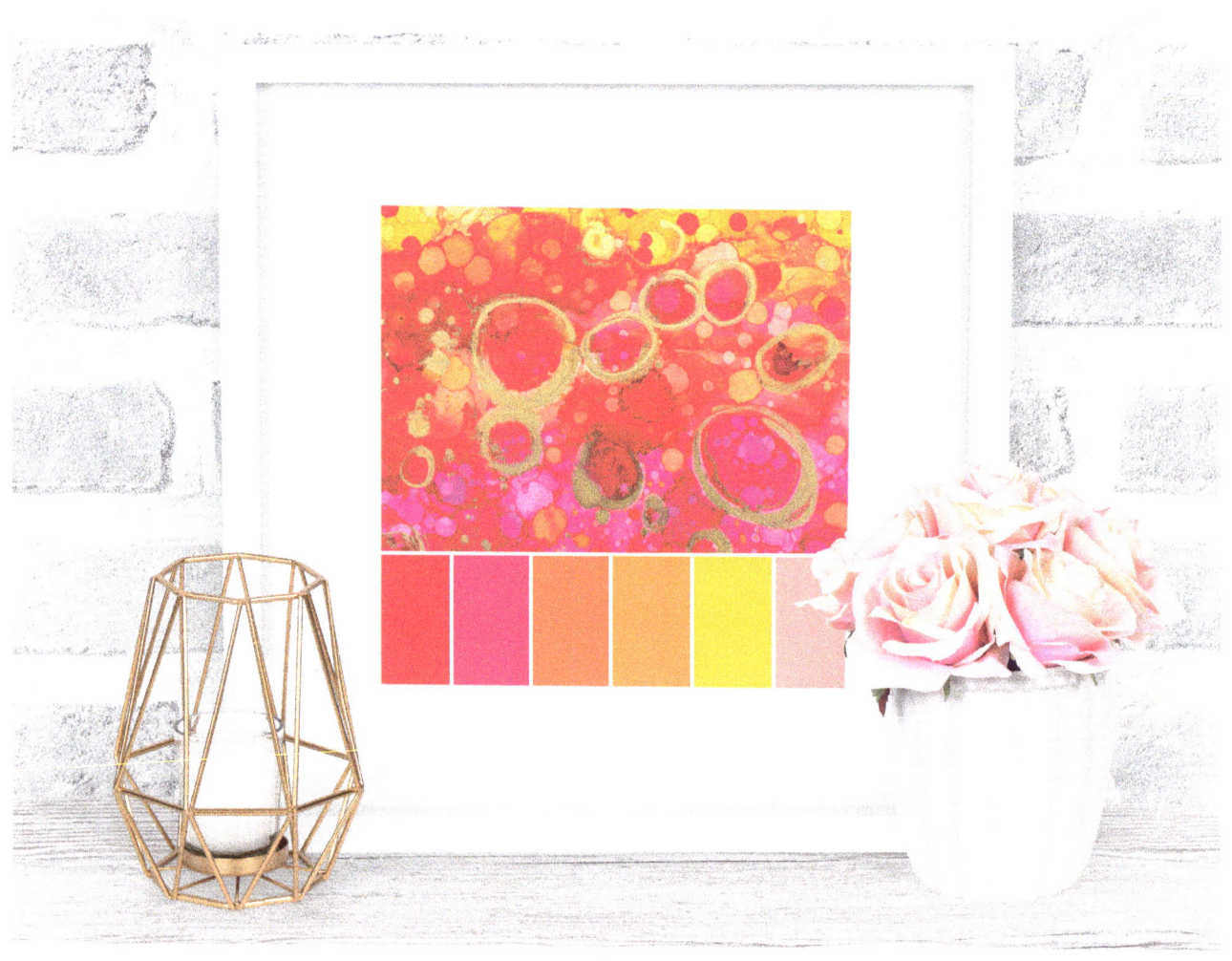

Once your painting has dried (alcohol inks dry very quickly),
you can frame your art, cut it up to make smaller art, such as cards,
or even mount your painting to a wood panel.

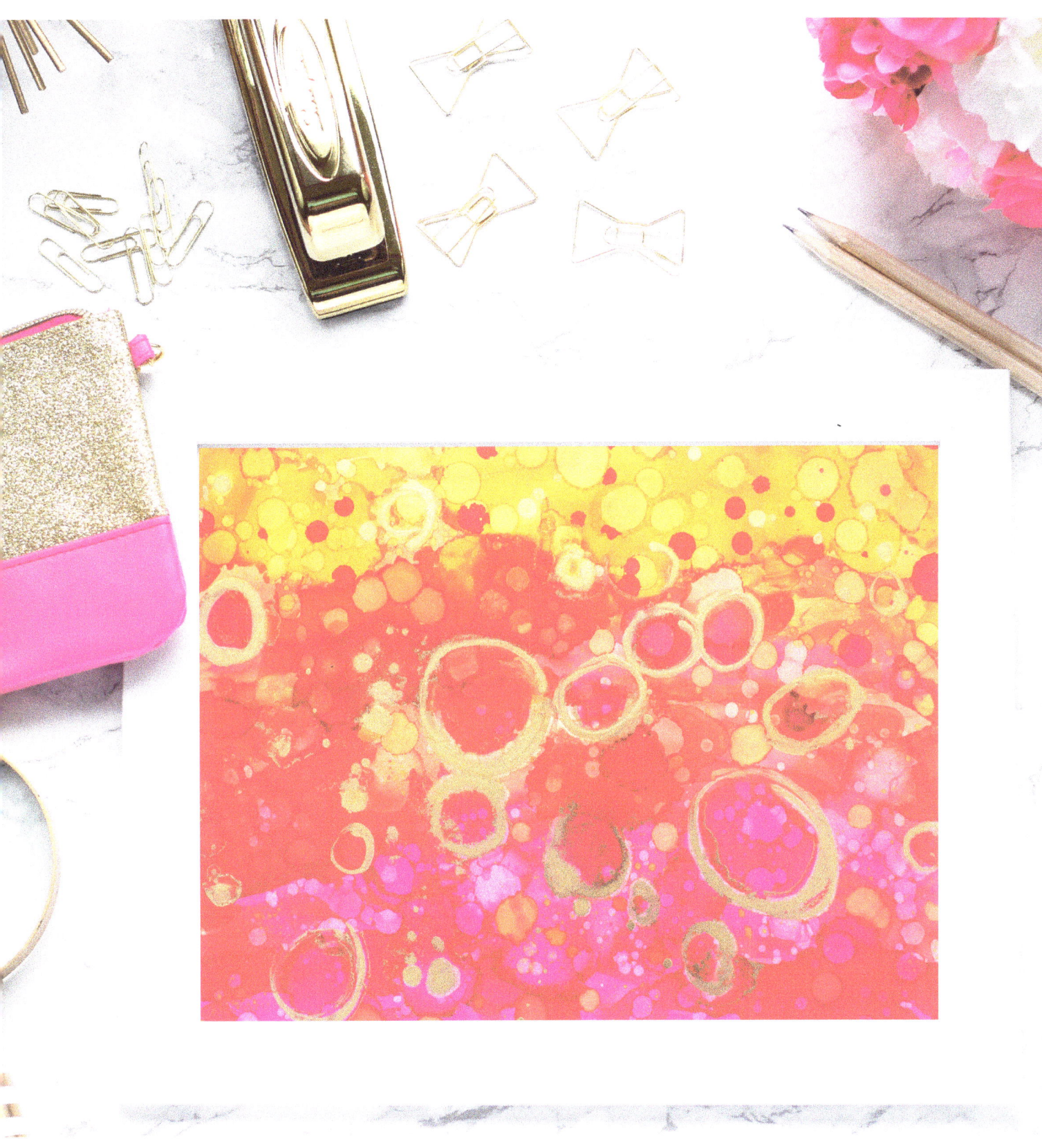

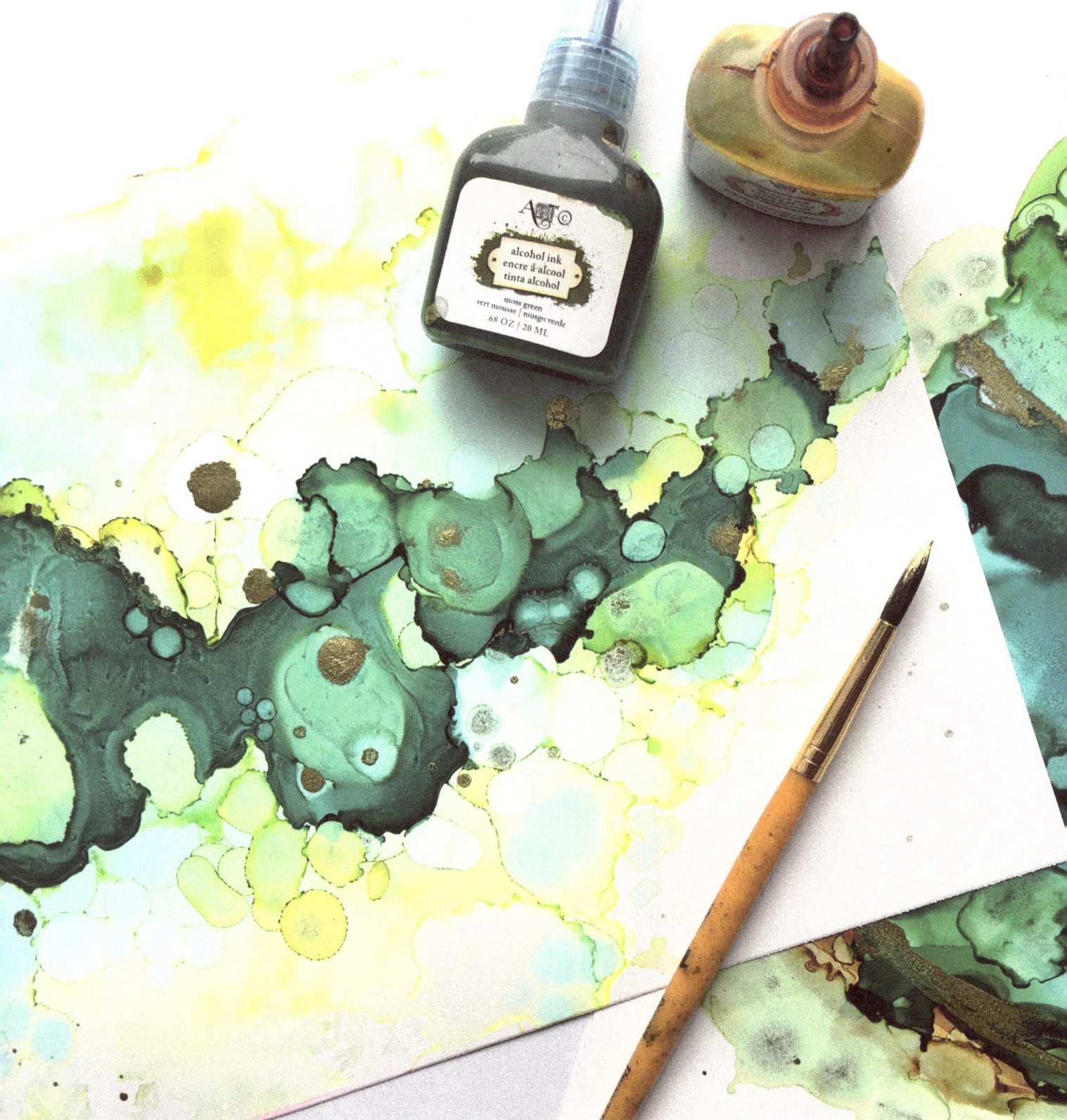

STYLE 2
SOFT AND MELLOW

In this style, you will create a more subdued, mottled look using blending solution and three alcohol ink colors.

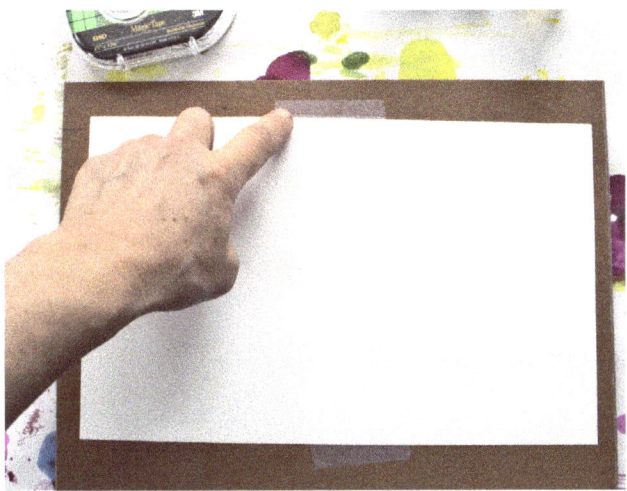

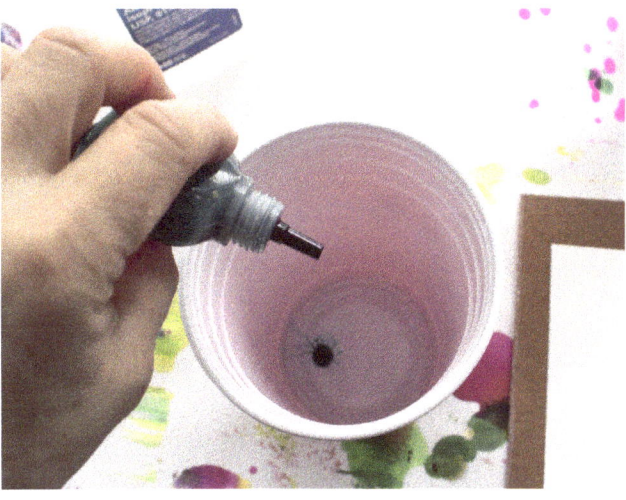

1. Start by taping your paper to something sturdy, such as a piece of cardboard or hard plastic. This will keep the paper flat and make it easy to move the page around when moving the inks across the surface.

2. Begin with some blending solution or isopropyl alcohol in a plastic container and add one drop of ink to the solution.

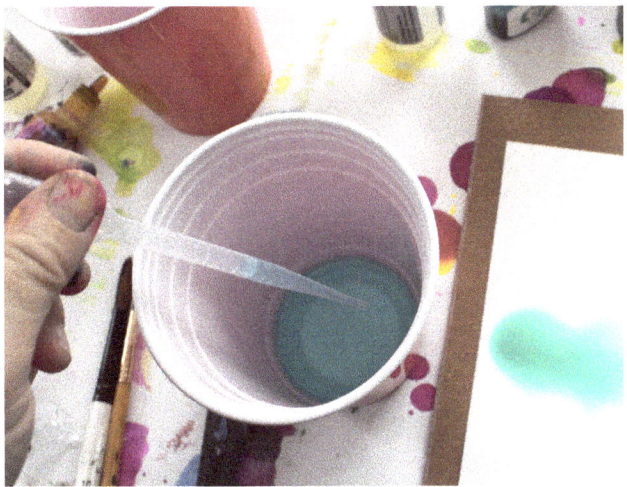

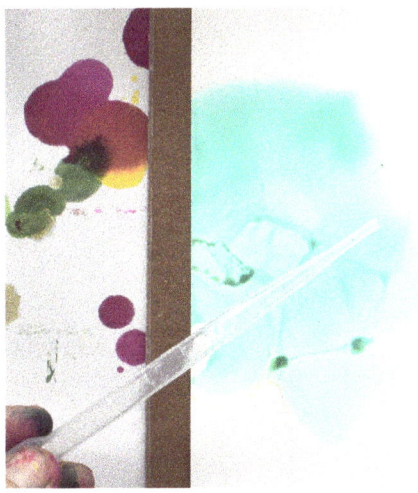

3. Swirl the cup around until the ink is incorporated into the solution. Use a dropper to draw up some ink and drop it onto the paper.

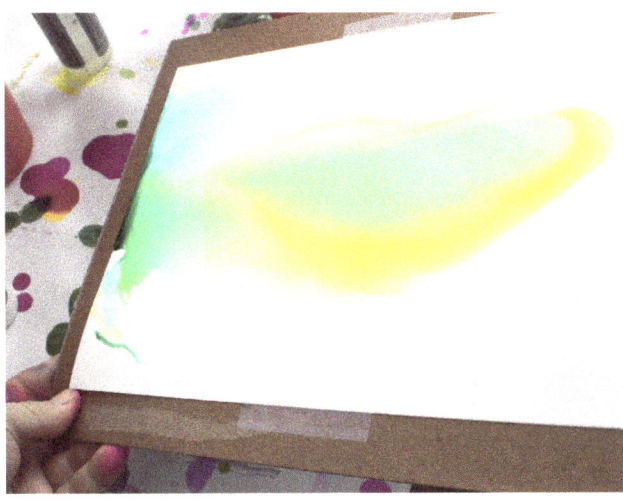

4. Pick up the board and tilt the paper around to move the ink with the solution. You may wish to add a second color to blend with the first. You can also use a hair dryer on the lowest setting to manipulate the inks (if the dryer setting is too high, you will have less control) or to speed the drying process. A heat gun is NOT recommended because the paper may warp or catch fire. (Alcohol inks are flammable.)

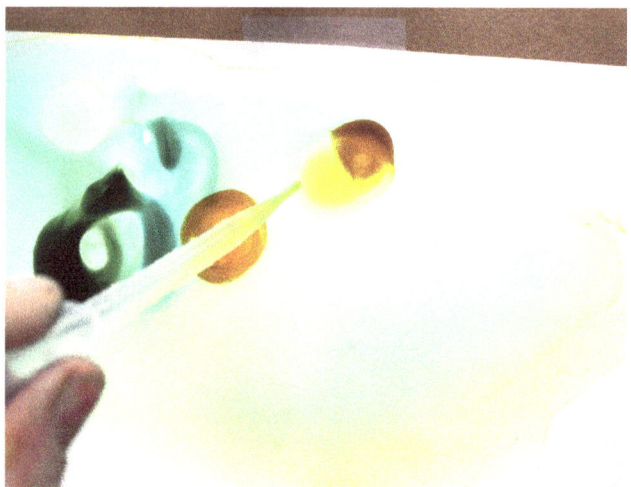 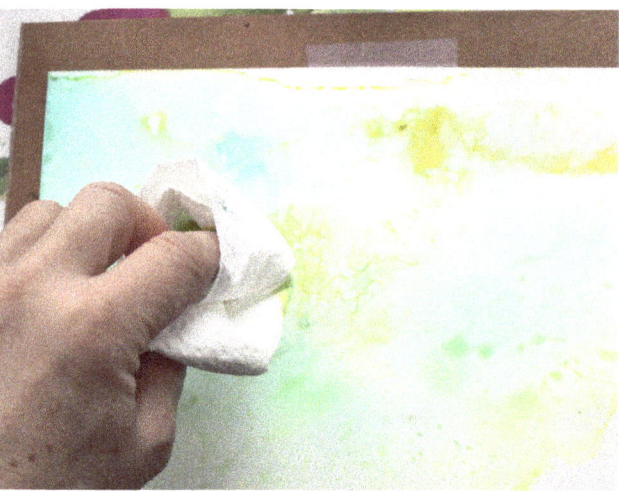

5. Add a few more drops of mixed ink solution to the paper. Then use a paper towel to lightly blot over the area to give it a textured look.

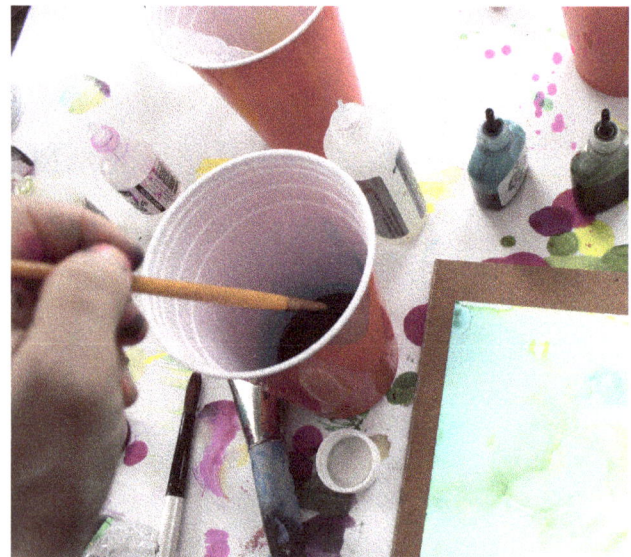 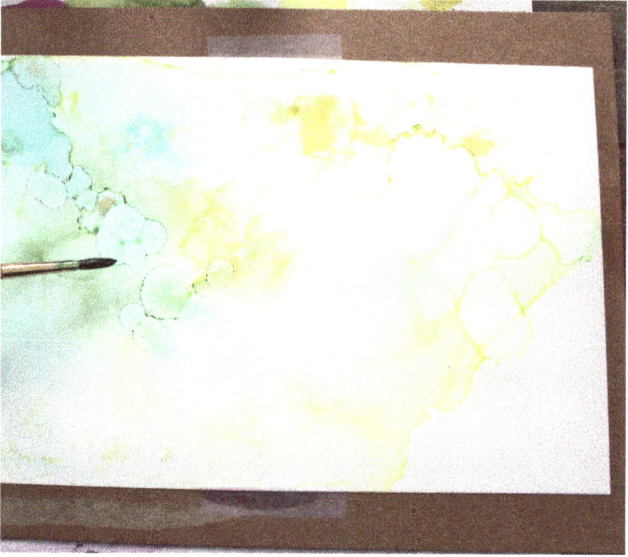

6. Darken the color by adding more drops of color to the ink solution mixture. Using a round brush for this step gives you more control. With the brush, lightly touch the ink and dab it onto the paper to transfer it. Continue applying the ink across the paper to create a second layer of texture.

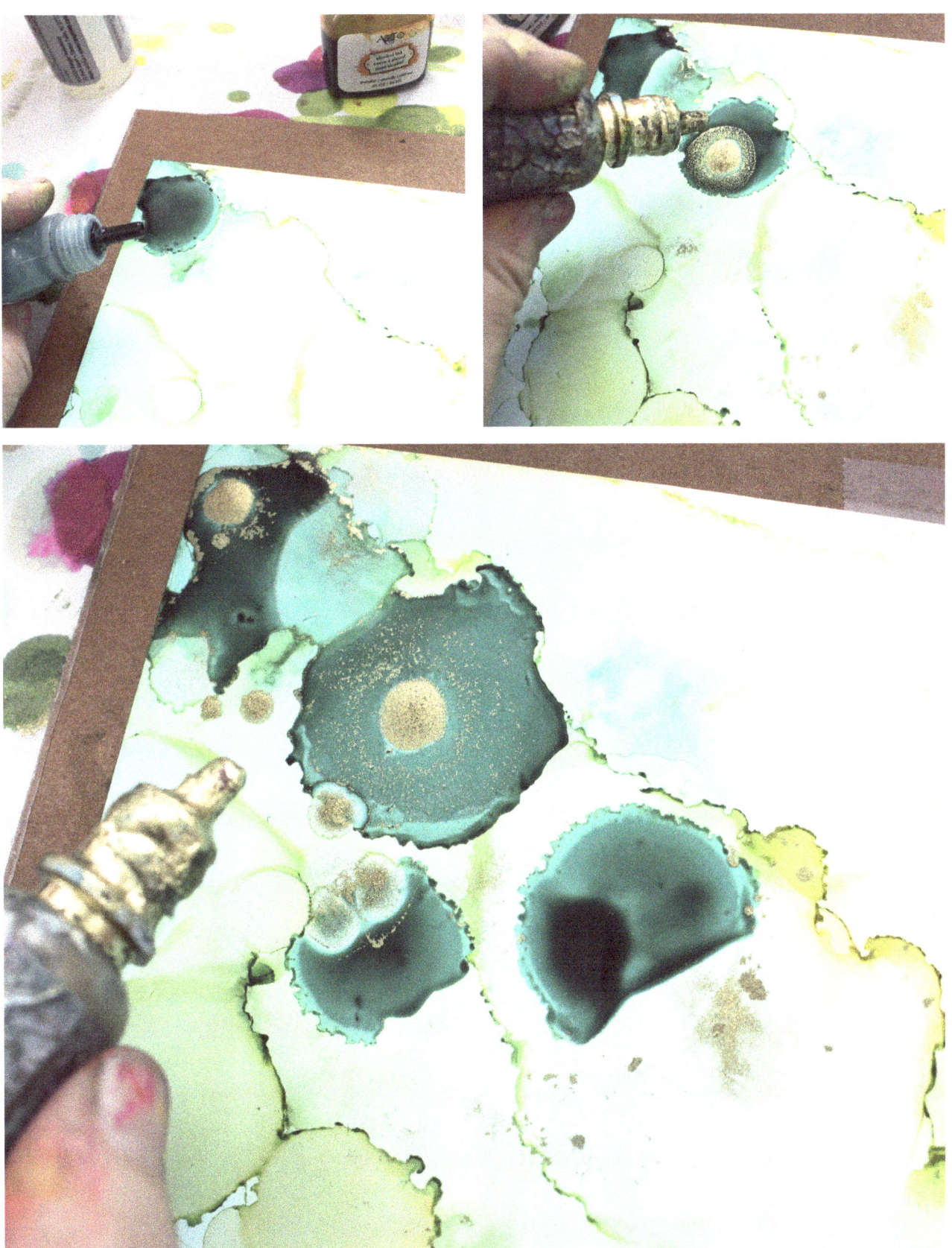

7. For the third layer, add a fully saturated ink color directly onto the paper, with a dropper, brush, or straight from the bottle. Continue adding drops and some rich gold metallic while the ink is still wet. The metallic will expand in the wet ink areas.

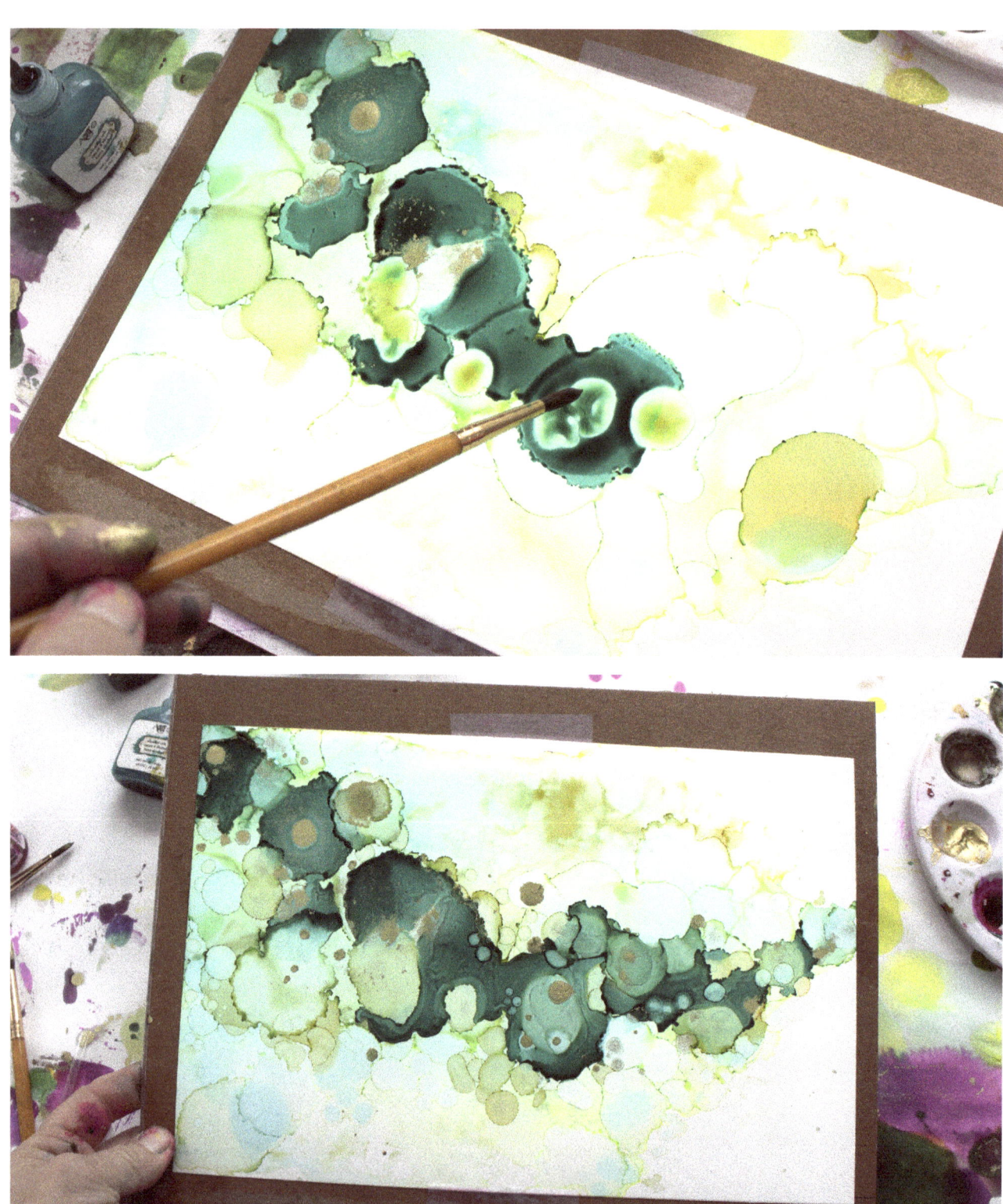

8. Continue with the process across the paper until you are happy with the result. Try to leave some negative space (white areas) in your painting. Turn your paper around as well to get a different perspective.

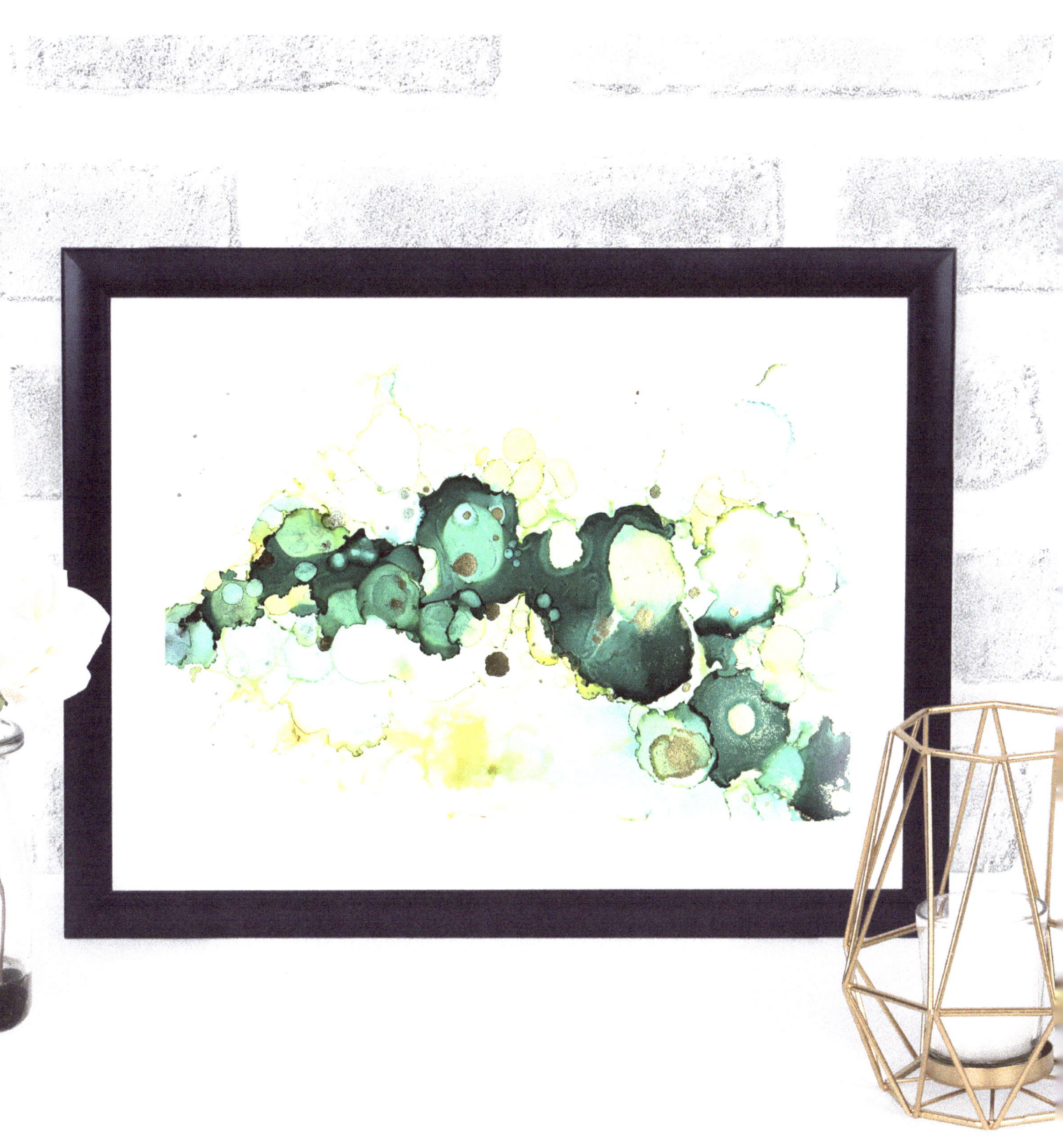

STYLE 3
BLOOMING ABSTRACT FLORALS

In this final alcohol ink painting, you will have fun with moving the inks around on your paper with either a straw or a can of compressed air to create an abstract floral look. The color combinations are endless, so choose a few favorites or start with one of the two versions shown here.

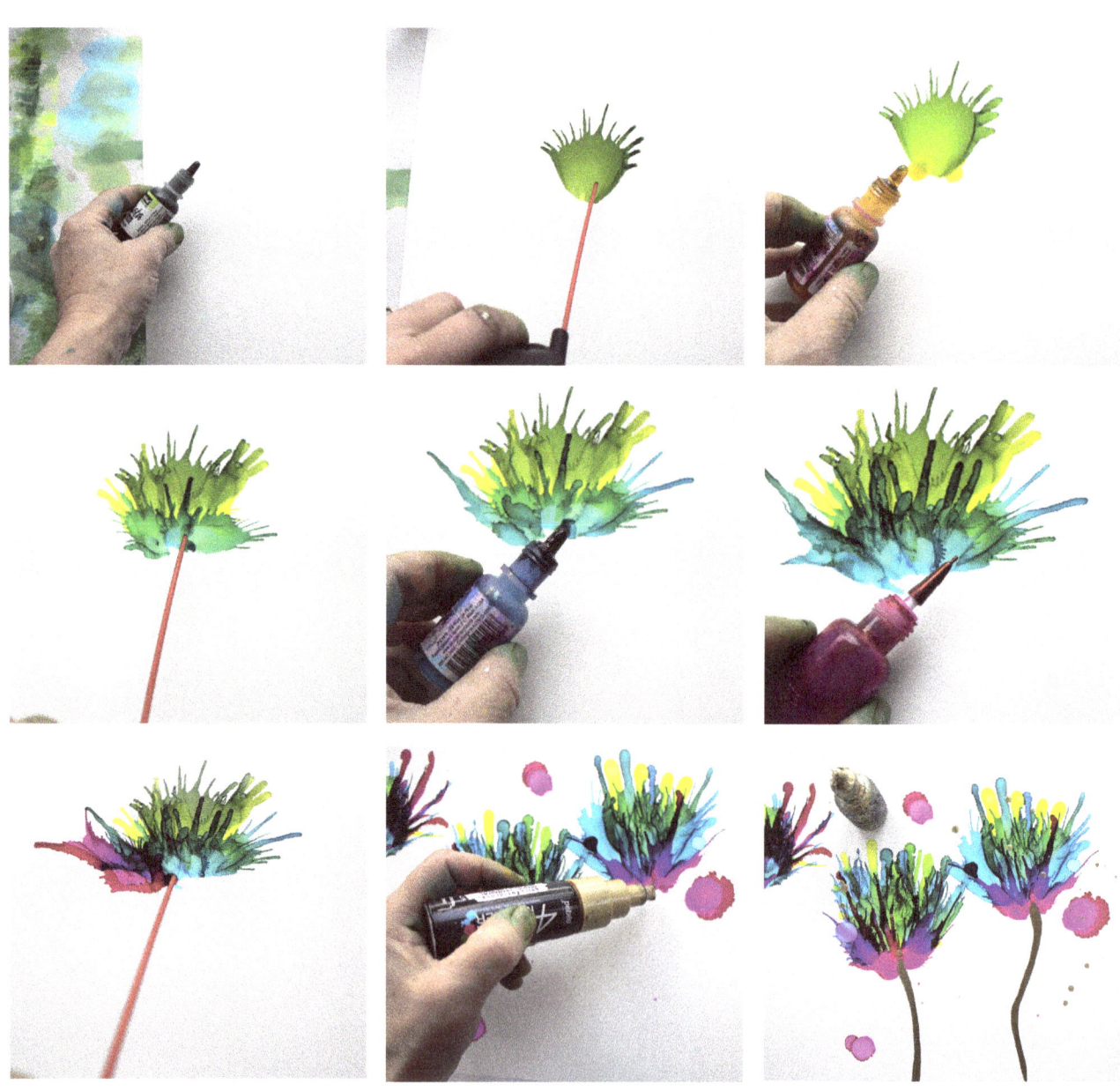

Floral 1: Lime Green, Sunbright Yellow, Baja Blue, Señorita Magenta by Jacquard

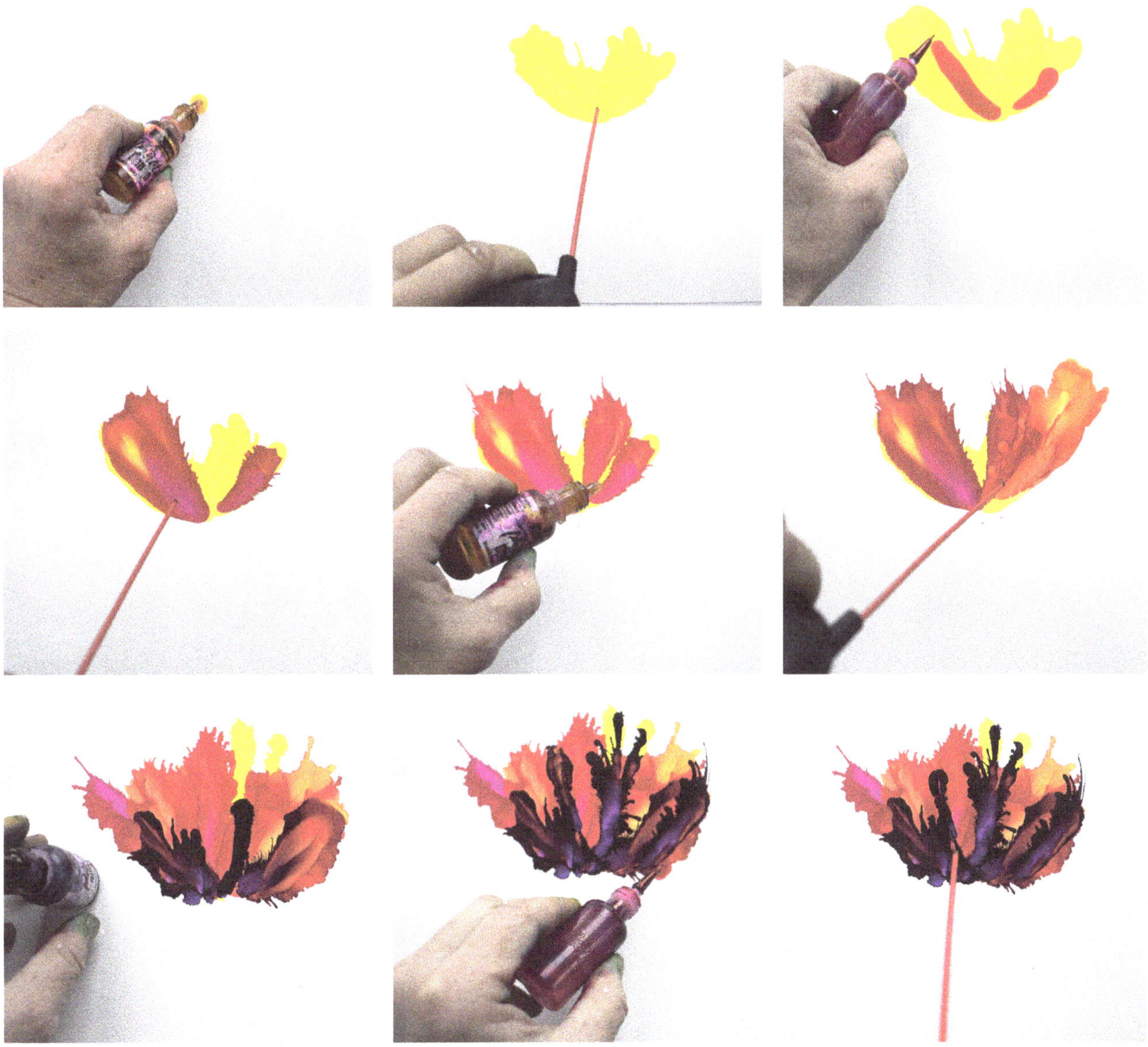

Floral 2: Sunbright Yellow, Señorita Magenta, Passion Purple by Jacquard

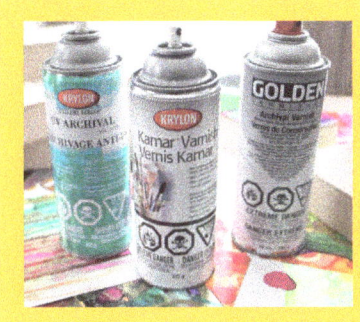

TIP

Once your paintings have fully dried, you may wish to frame them. Before doing so, you should consider sealing your art to protect it from fading so that it will last for years to come. Seal first with a spray product such as Krylon Kamar. This protects the inks and prevents them from bleeding. Then apply a UV protective coating such as Krylon UV Archival or Golden UV Varnish. Allow coats to dry between layers.

VARIATION IDEAS

Continue to experiment with alcohol inks and you'll find there are many other ways to manipulate them. Here are just a few ideas. Have fun and experiment!

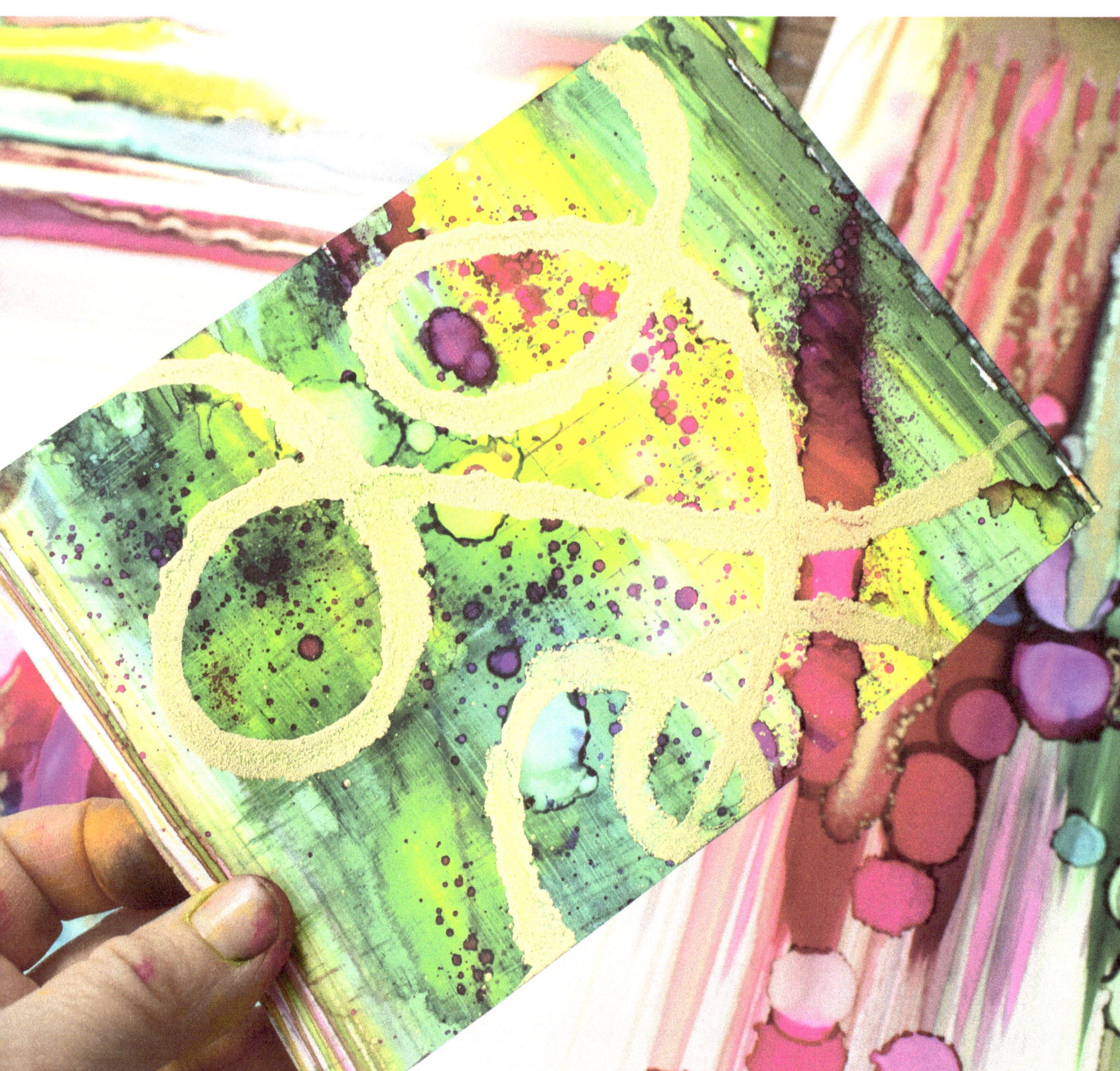

Results from multiple techniques of dropping ink, swiping, brushstrokes, and embellishing with metallic gold

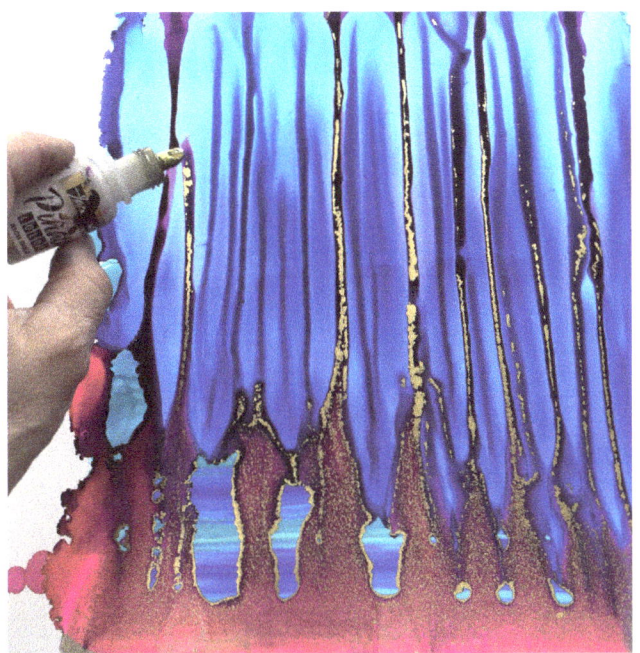

Results from lifting paper and allowing multiple colored inks and blending solution to run in one direction up/down the page

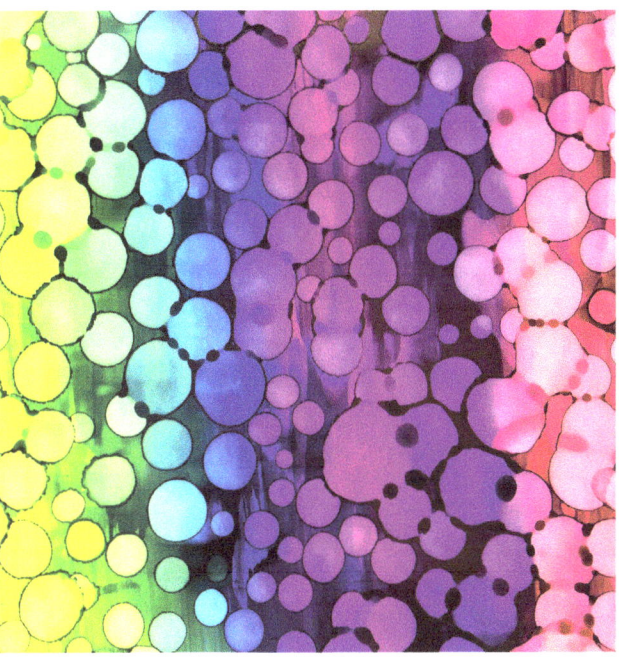

Circle results from sponge-brushed color and drops of alcohol or cleanup solution on paper

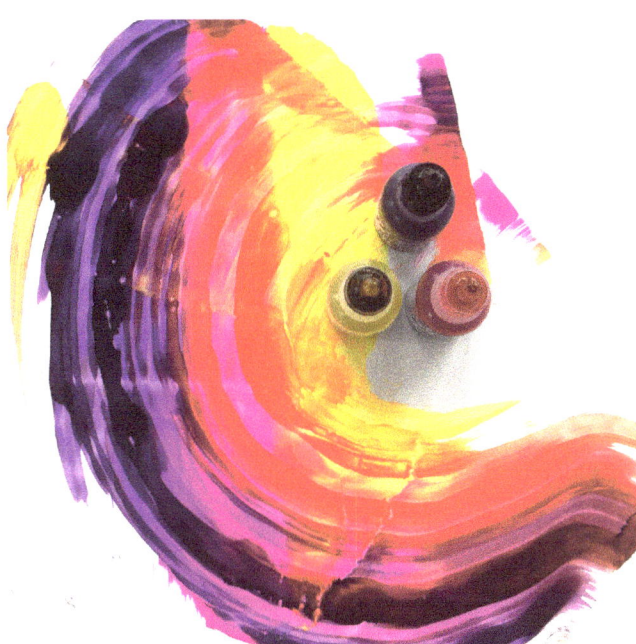

Results from using a plastic ruler to swipe color in a circular motion across the paper

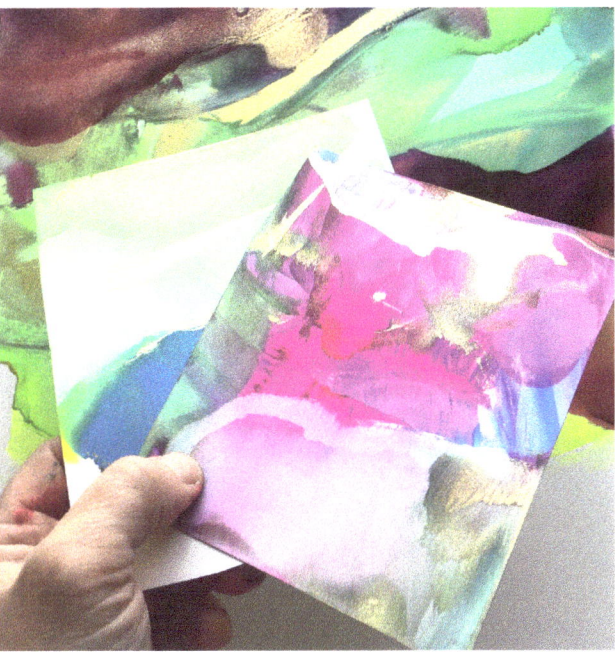

Results from swiping two pieces of Yupo paper together

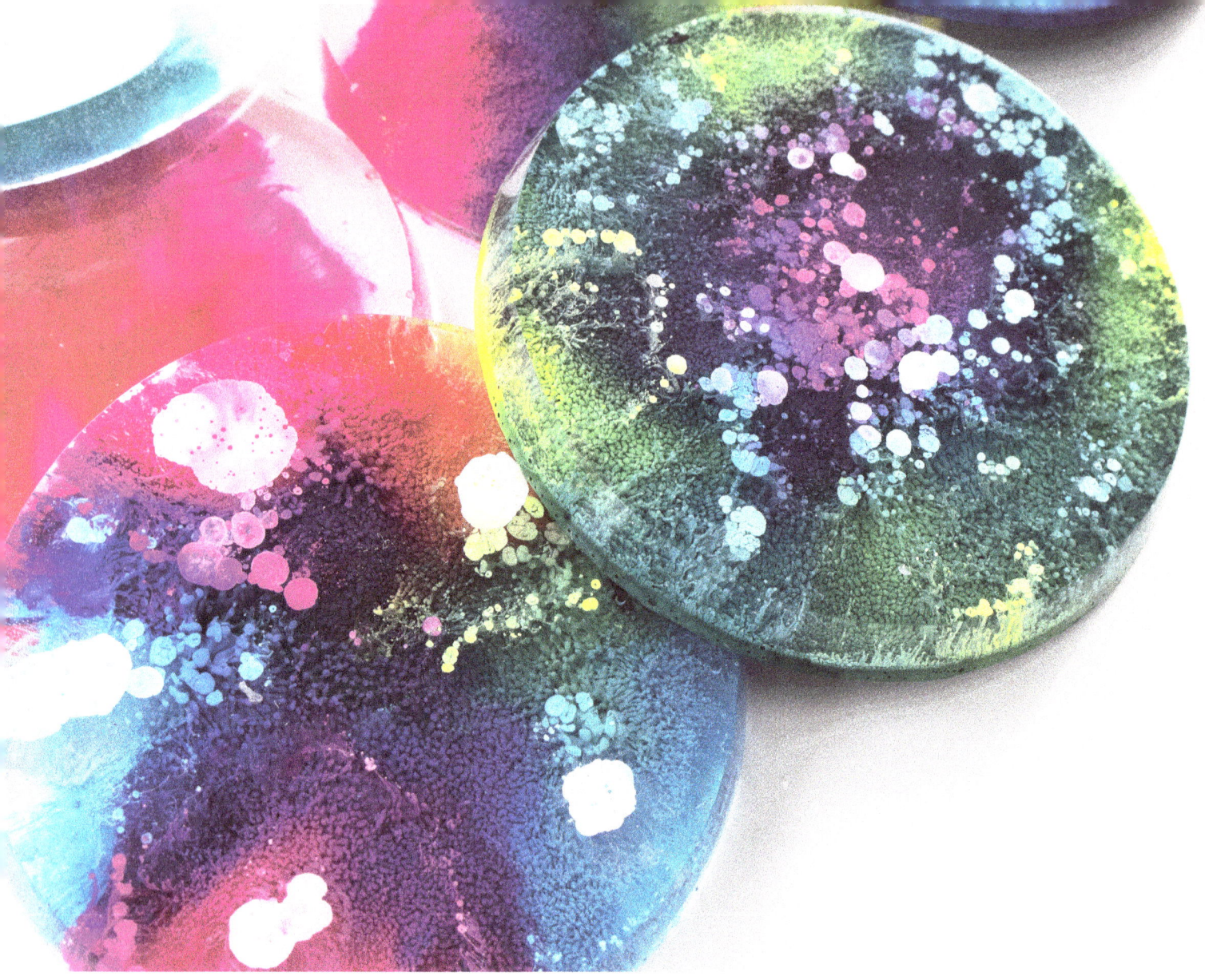

ALCOHOL INK AND RESIN PETRI MOLDS

Resin petri molds exploded on the scene a few years ago and have since become a phenomenon. How they started is anyone's guess, but there are lots of artists and hobbyists making these eye-catching pieces of fluid art. They are so much fun to create and the outcome is different every time, which makes them appealing and addictive. The main component in these designs is resin. If you haven't used resin in your art projects, you're missing out. Resin is like liquid glass and easy to use. Resin adds depth and can be colored several different ways, but the main colorant used in this project is alcohol ink. You can turn petri molds into coasters and refrigerator magnets, frame them in a shadow box, or put them on a stand. No matter what, you'll want to make lots of them!

MATERIALS

black garbage bag or plastic tablecloth to protect surface area

protective gloves

safety mask

clear plastic measuring cup

large craft sticks for stirring

resin kit*

round silicone mold

mini butane torch

alcohol inks, including white

dropper tool

toothpicks

silicone and/or isopropyl alcohol

container to cover mold

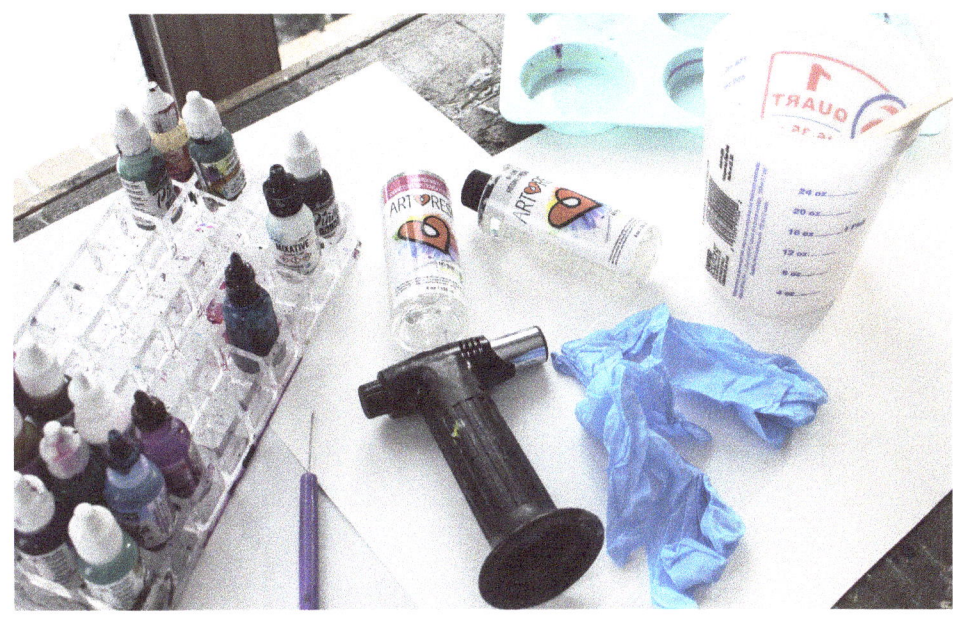

*Note: Different brands of resin have various working times (i.e., the time before it starts curing). Choose one such as Art Resin, which has a working time of about 45 minutes.

TIPS

- Make sure you use a glossy bottom mold or your petri resins will not be shiny when you remove them from the mold.
- Use a brand of resin that contains zero VOCs (meaning they are nearly odorless), is made for art, and is safe for indoor use.
- Use a butane torch to eliminate bubbles.

Work area covered with plastic sheeting

SAFETY

- Work in a well-ventilated area with a mask if necessary.
- Wear gloves to protect your hands—resin is VERY messy and sticky.
- Cover your work area with plastic tablecloths, parchment paper, or black garbage bags to prevent any dripped resin from permanently curing to your tabletops or work spaces.
- **Warning:** Do NOT torch heavily over alcohol inks, as they are flammable.

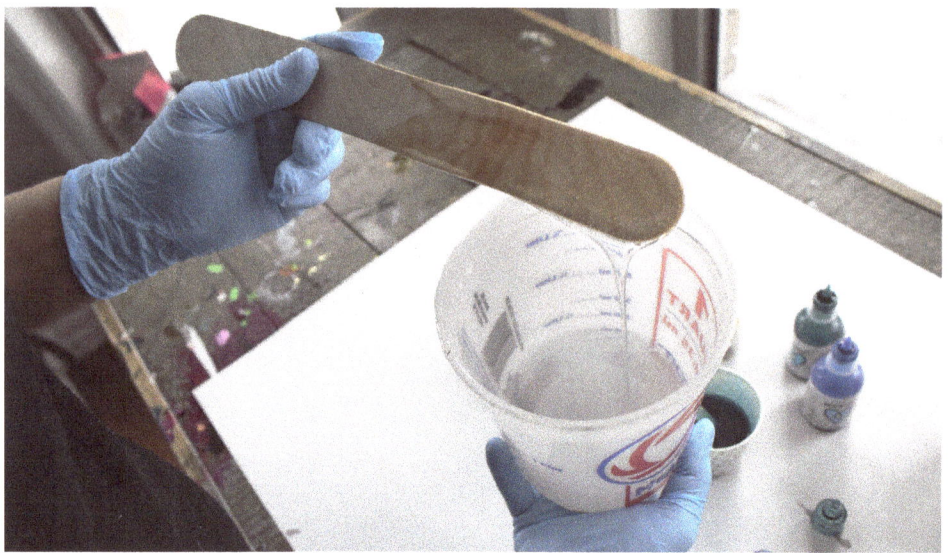

1. Cover your work area with a large plastic bag. Wearing safety gloves and a face mask, use a clear plastic measuring cup and a craft stick to mix the resin. Most resin kits are a two-part epoxy system in which you mix equal parts of both the resin and the hardener.

TIP

Using a large craft stick, mix the resin slowly for 3 minutes (or whatever your resin brand requires) to avoid excess bubbles. Work in a room temperature of at least 72°F (22°C). You can also place your mixed resin container in warm water to help speed the release of bubbles. (Don't get any water in your resin, as it will turn cloudy.) Mix thoroughly and scrape the bottom and sides of the container.

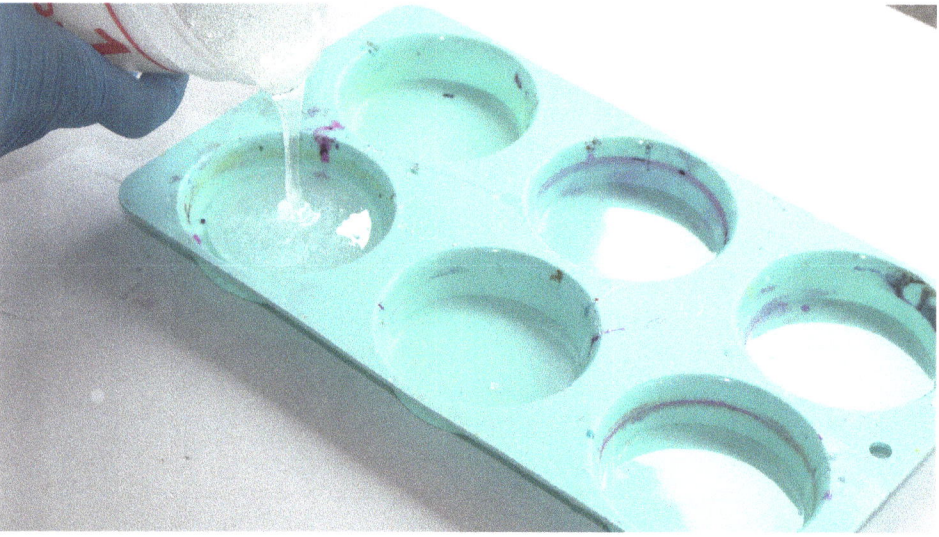

2. Once the resin has been mixed thoroughly, slowly begin pouring equal amounts of resin into each mold cavity. Don't fill your molds to the top; if there is too much resin in the mold it may not cure properly. Limit each resin layer pour to ⅛" (3 mm) thickness.

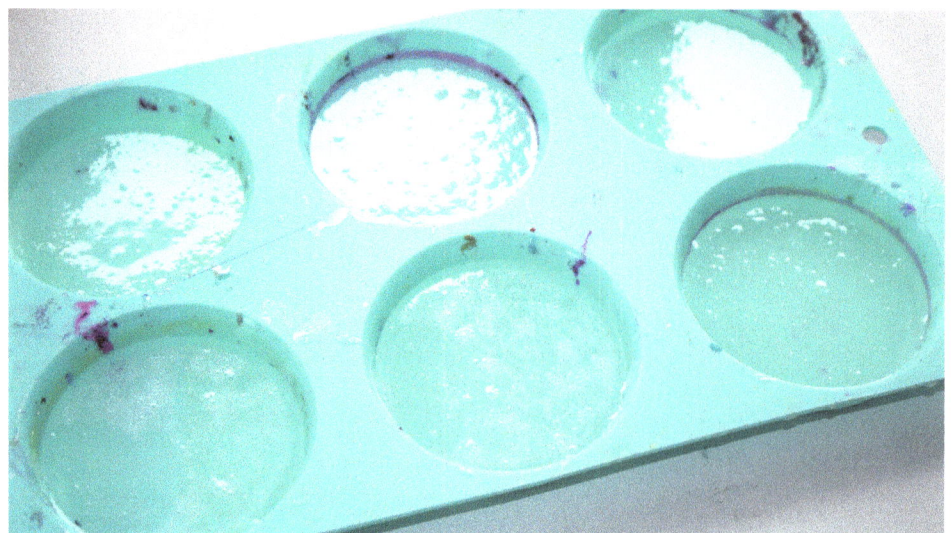

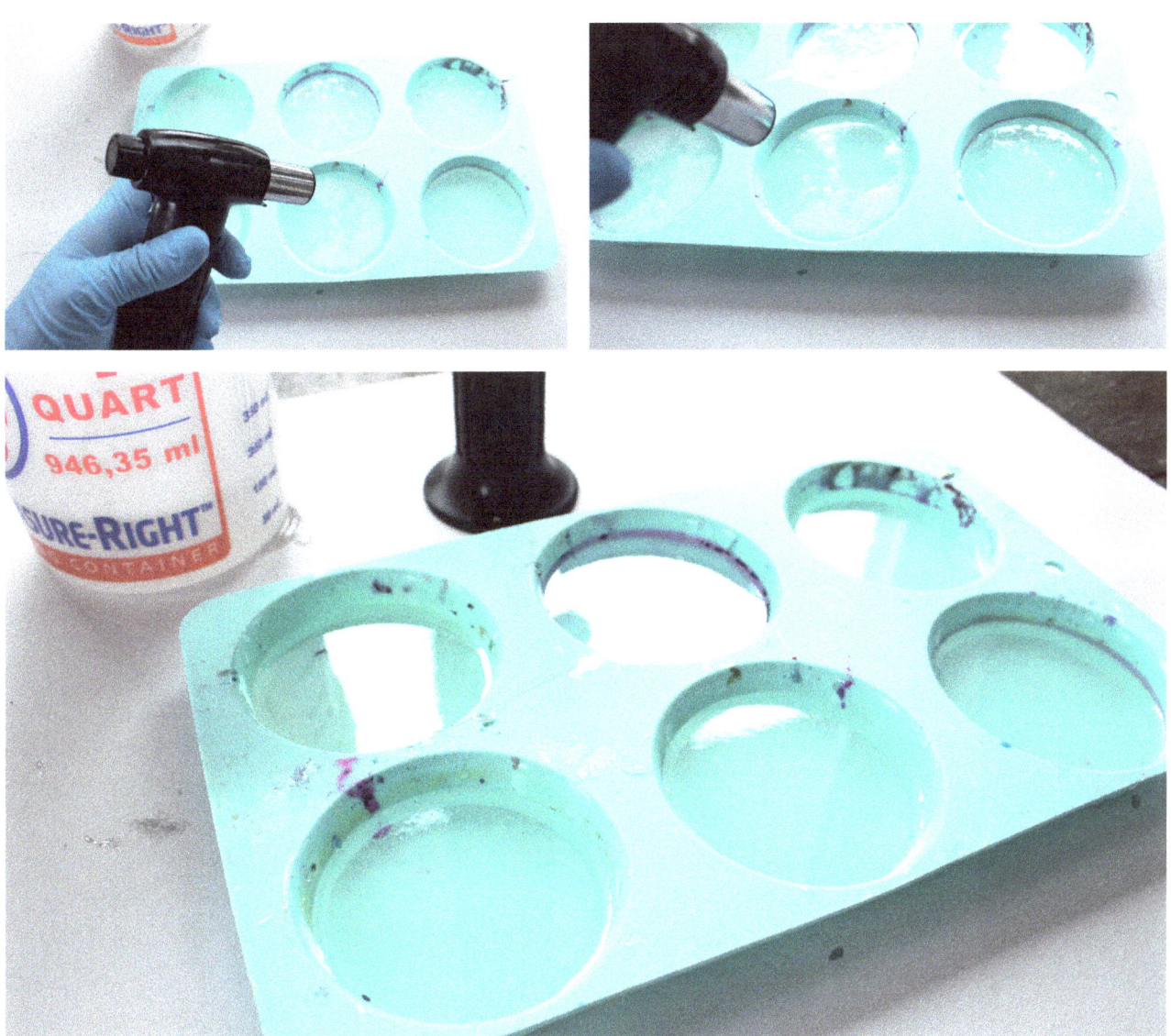

3. Move the torch lightly over the resin in quick circular motions. Allow the flame to just barely kiss the surface and do not hesitate too long over one area. Stop when your resin becomes clear on the surface.

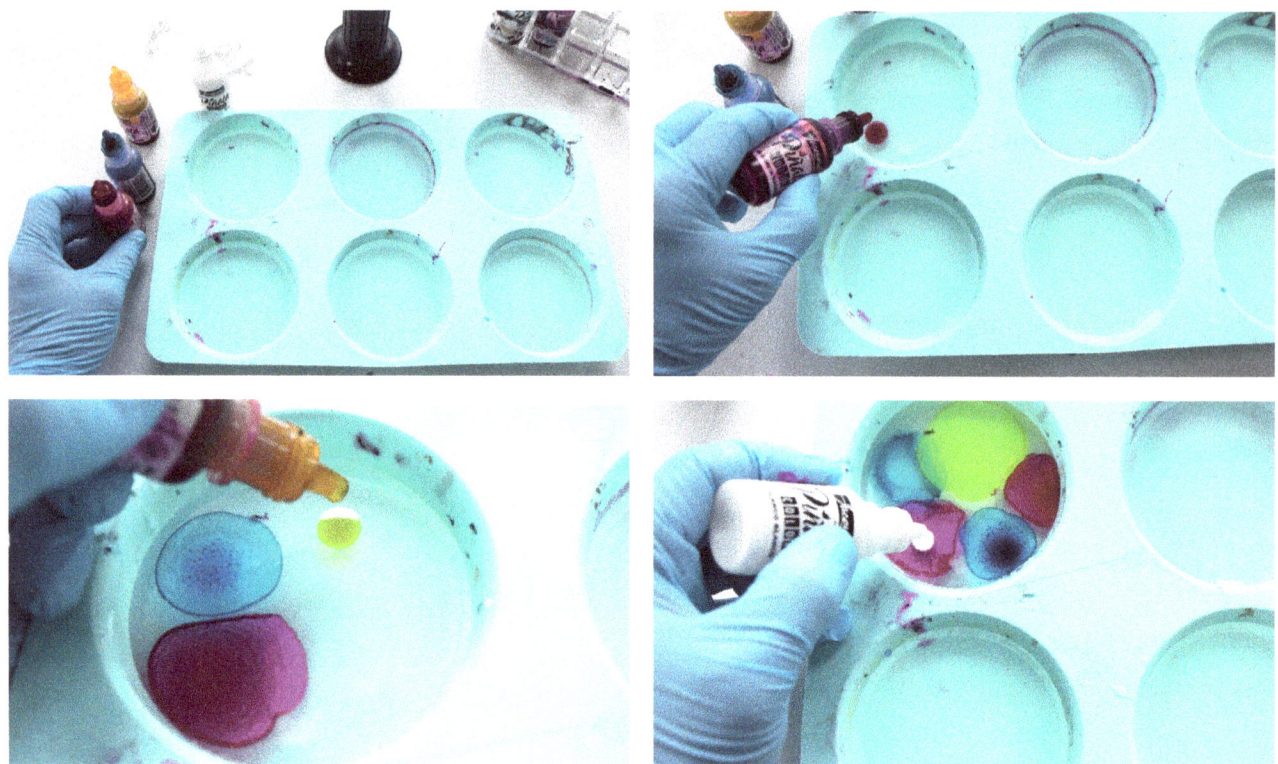

4. Select several alcohol ink colors, including white. The white color is pigment-based and will help push the other inks down into the resin. Begin adding individual drops into the resin. Use a dropper tool if you don't want to squeeze straight from the bottle.

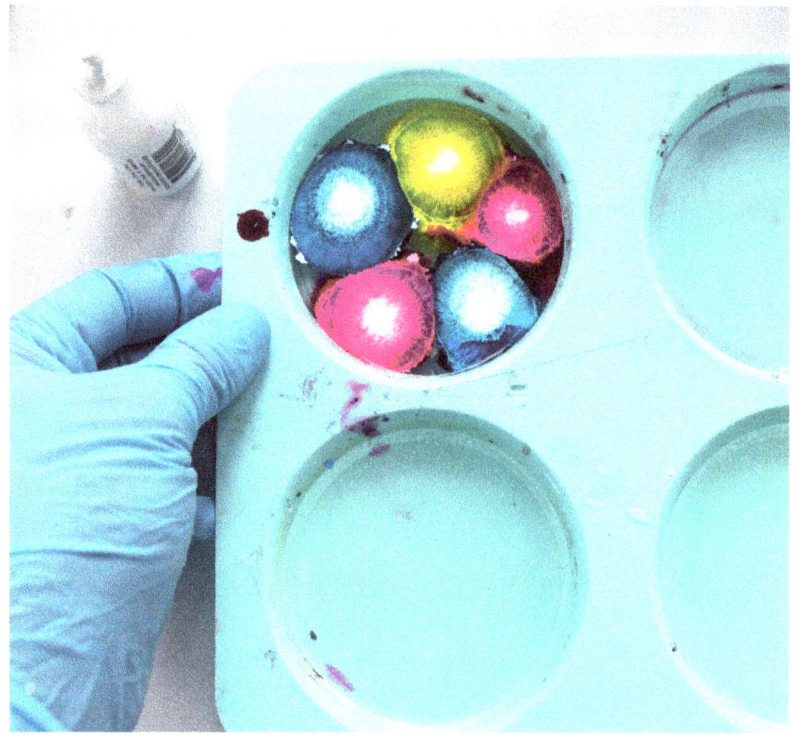

5. Continue with one drop of ink followed by one drop of white around the mold. Do this three or four times. The inks will start to react and change before your eyes.

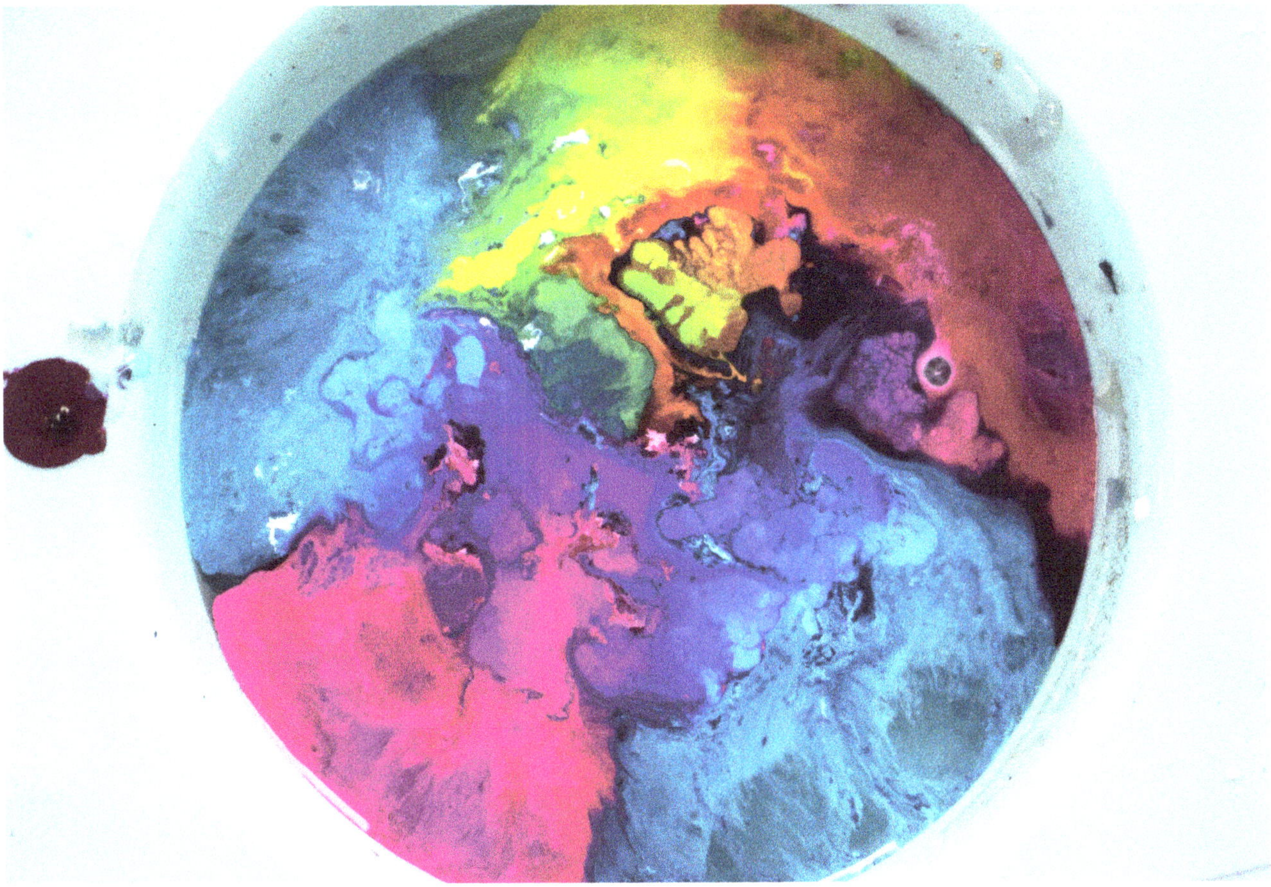

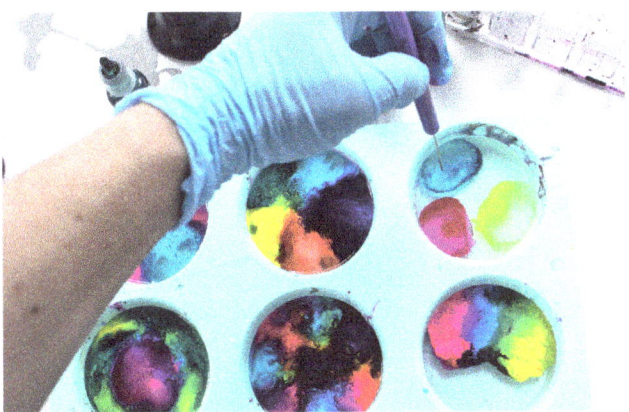

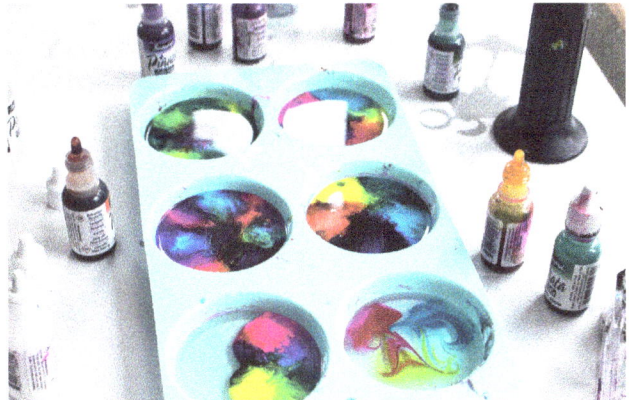

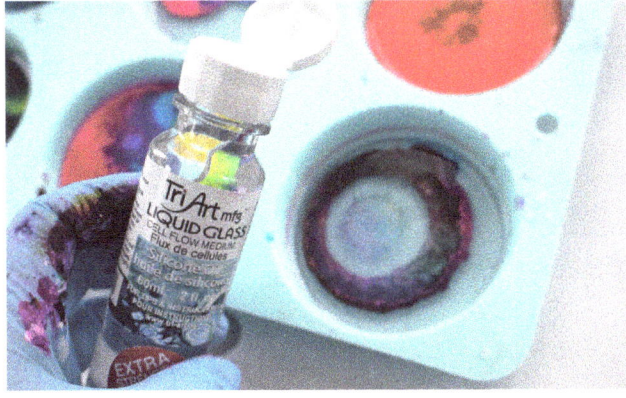

6. You may also want to try adding just a few drops of color in one area while leaving negative space in another part of the mold. Use a toothpick to lightly swirl the inks to get a different effect. You can also add a drop of silicone or isopropyl alcohol to achieve various results. Silicone and resin do not mix, so your inks will push away when the silicone is dropped into the mold. If using silicone, only use a drop or two because it may affect the resin and not cure properly.

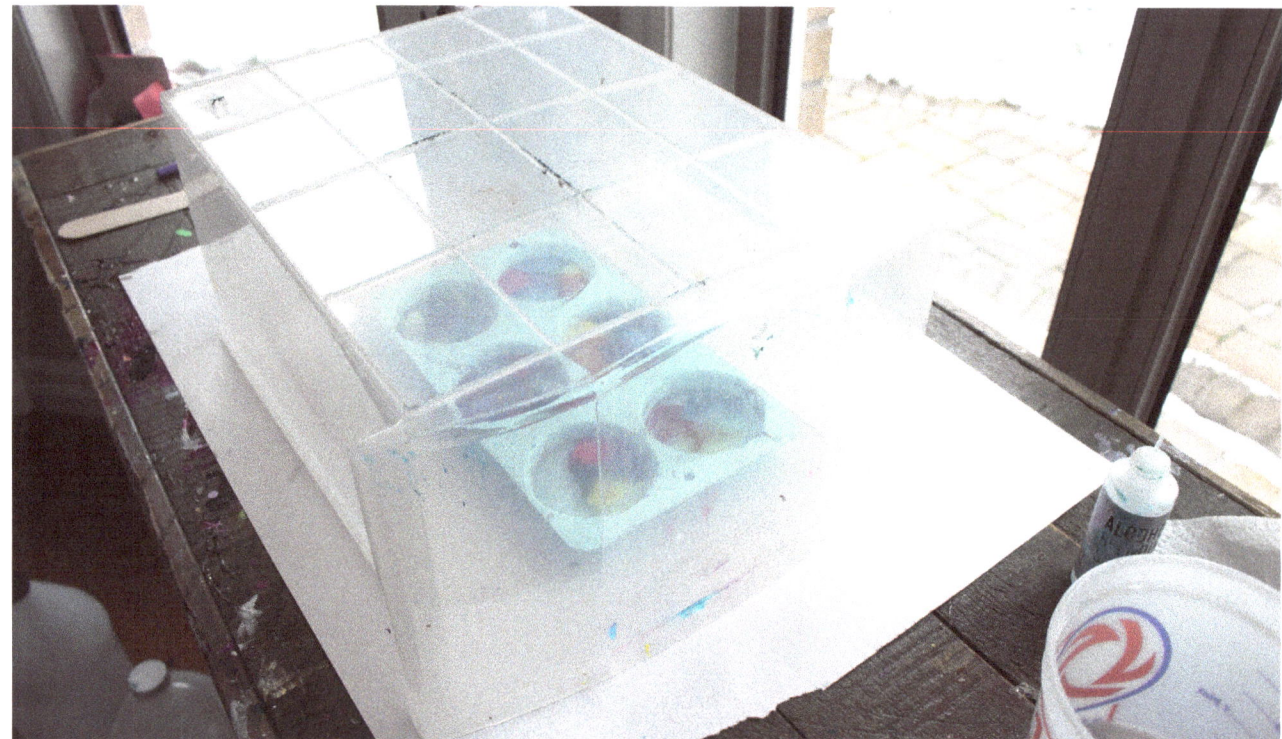

7. Once you feel your design is complete, cover the entire silicone tray with a container to avoid dust from settling into your resin while it cures. Leave your petri molds for at least 10 hours before removal. Resin is hard and fully cured after 24 hours.

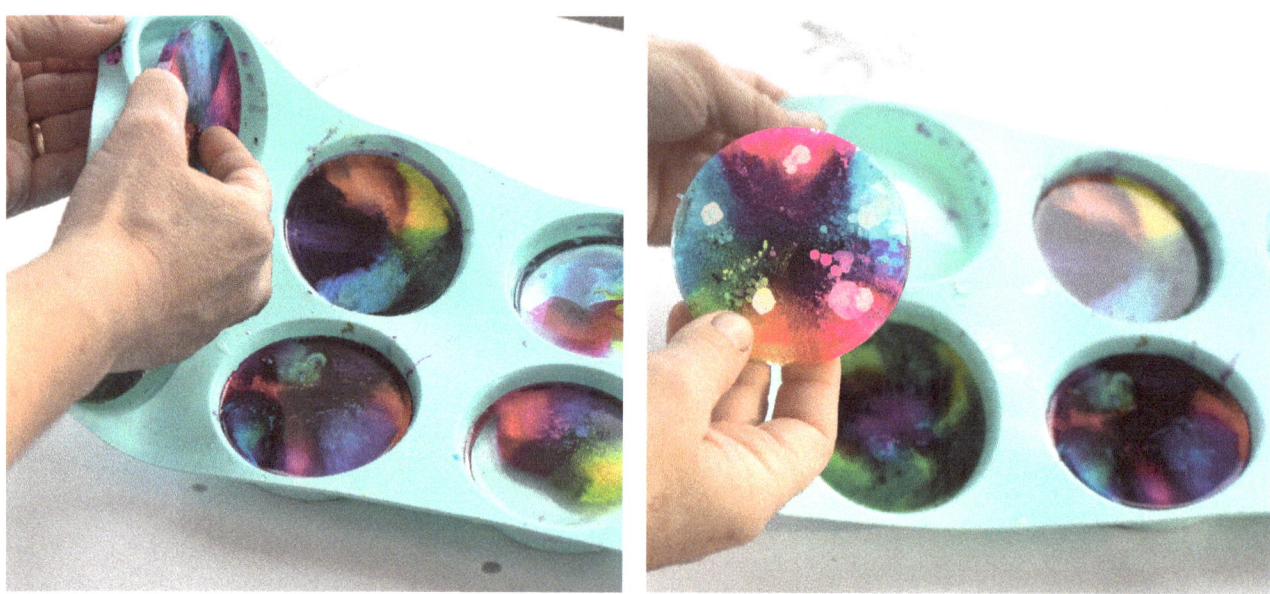

8. Slowly peel back and pop out the almost fully cured resin from your molds.

TIP

Remove petri resins from your molds between 10 and 15 hours after pouring, when they are still slightly soft. This prevents resin from sticking to the silicone mold, which can cause your mold to tear.

Admire the colors and results! Place them on a small stand, mount them in a frame, or just place them where they will make you smile.

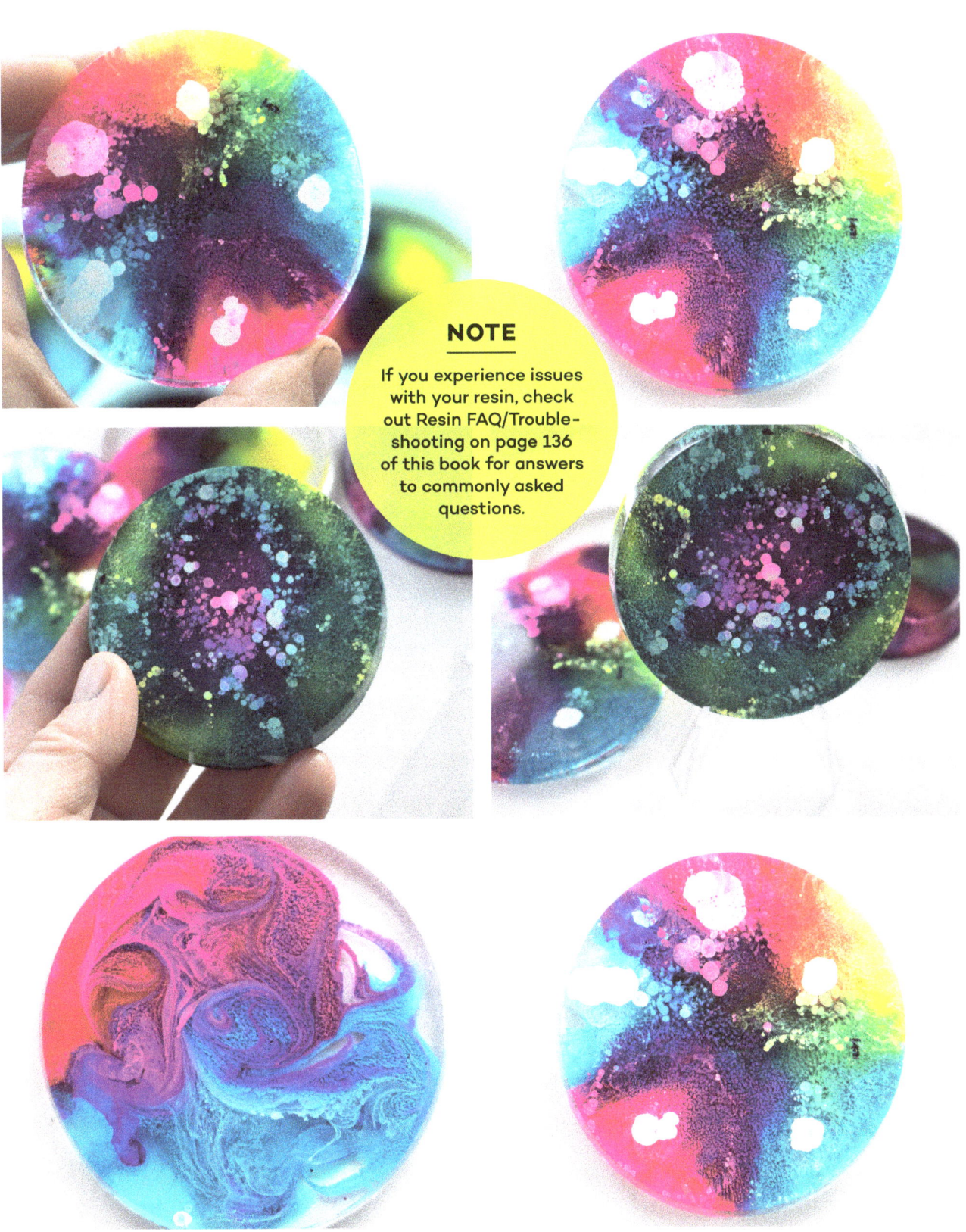

NOTE

If you experience issues with your resin, check out Resin FAQ/Trouble-shooting on page 136 of this book for answers to commonly asked questions.

MOD MINI ALCOHOL INK AND RESIN COLLAGE

"MOD Mini" is a name I created after brainstorming ideas for my miniature alcohol ink collage art, displayed on wood panels and finished with resin. MOD is short for "modern" and mini is its size, all small 4-inch (10 cm) squares. These mini artworks look like colorful jewels. They make for a wonderful wall collection or look great sitting on your shelf or mantle. They became an instant hit when I began posting them online and I have since created an online course to teach people how to make their own. Now you can create these fun little pieces of art made from Yupo scraps, some adhesive, wood blocks, and resin. In the first stage of the project, you'll adhere your artwork to the block and seal it. In the second stage, you'll apply the resin.

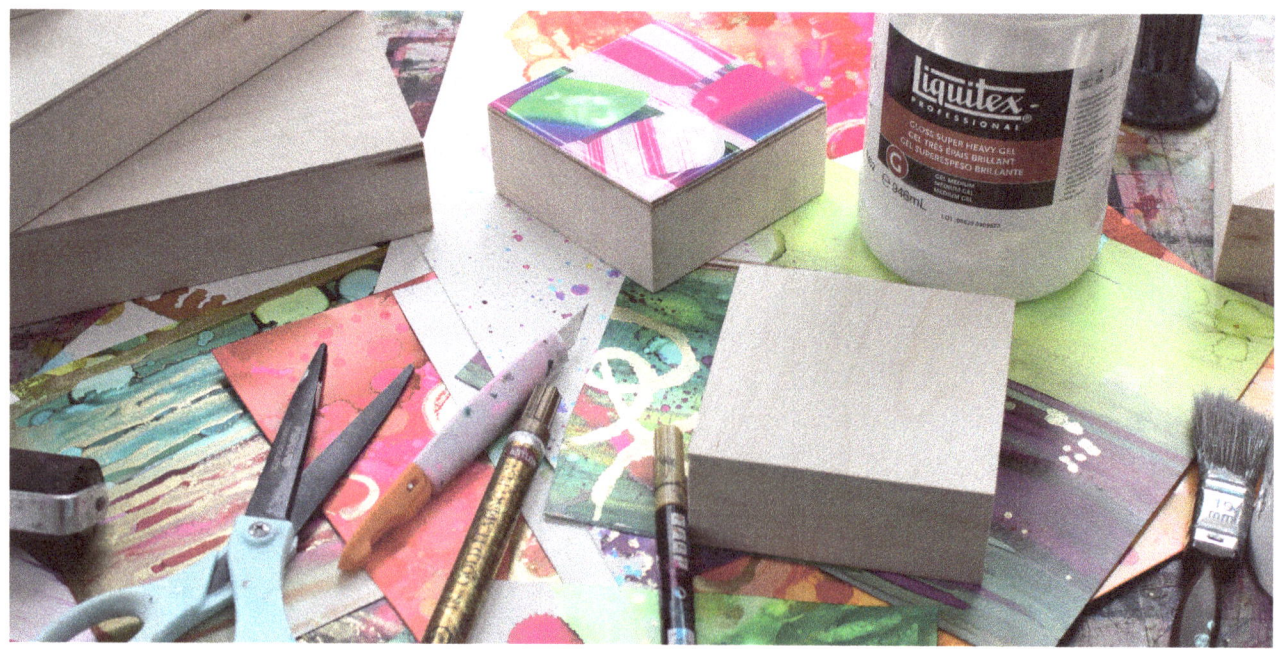

WOOD BLOCK STAGE

MATERIALS

Yupo painting remnants or scraps

a small-sized artist wood panel

wood sealer

old paintbrush

medium body matte or glossy gel, spray adhesive, or YES Paste

X-Acto knife

embellishing metallic pens/markers

glue sticks, thin-tip glue pens, or broad glue markers

painter's tape

safety mask

cardboard box

Krylon sealant

Krylon UV protectant spray

TIP

Find the online step-by-step course at www.janes classroom.com.

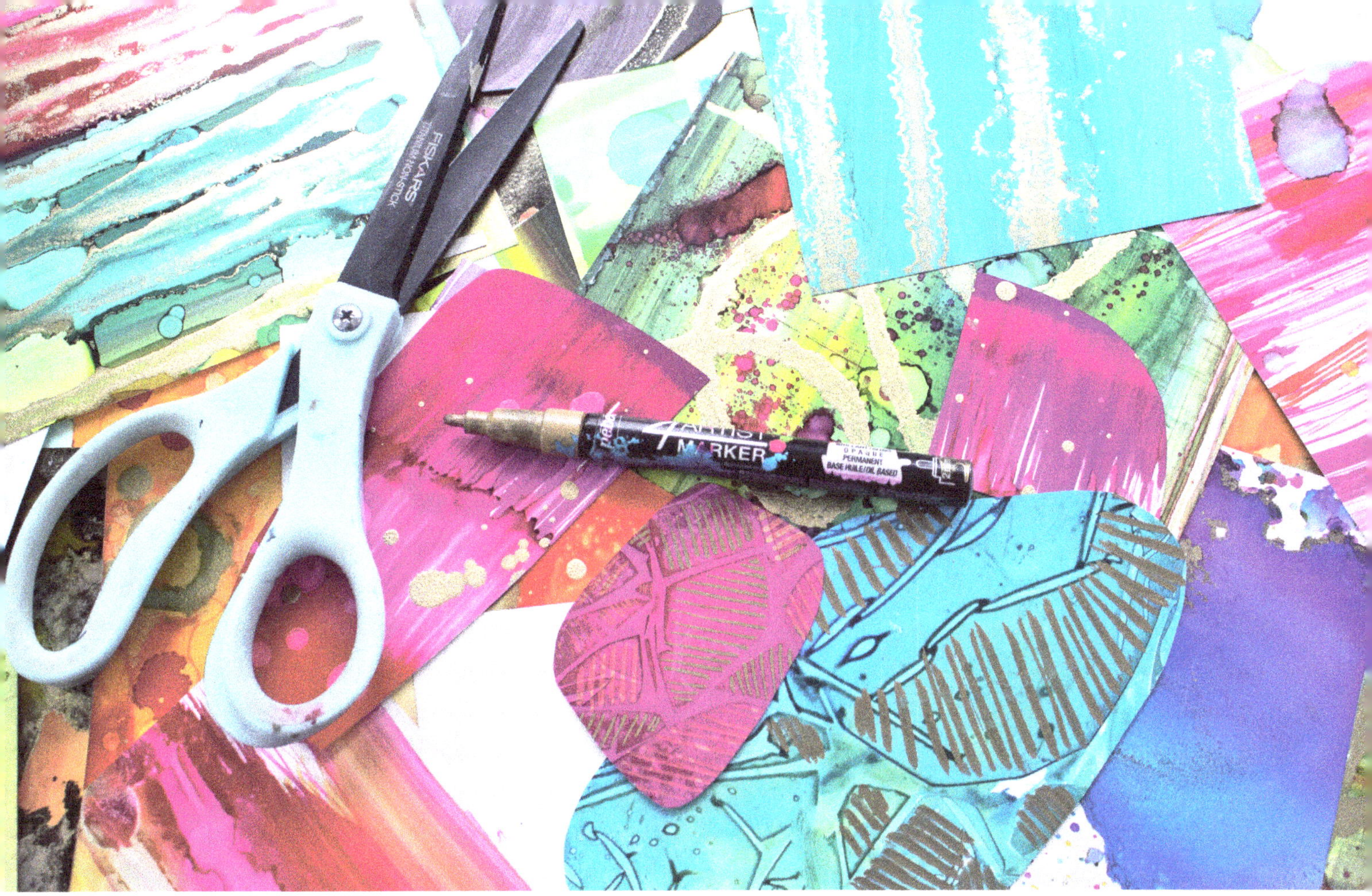

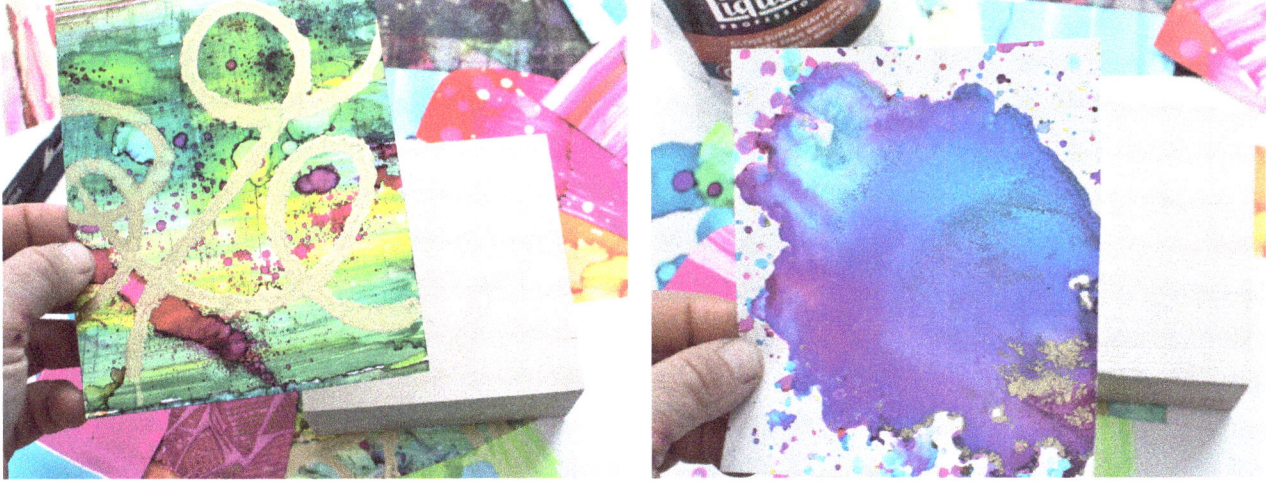

1. Begin by sorting through and selecting some colorful alcohol ink pieces of Yupo that you created from the first project in the book. You may have thought some of your experiments didn't amount to much or you didn't like how a few of them turned out. But never throw anything away! You can always repurpose these works and use them to create some very pretty color combinations, textures, and layers for your MOD Minis. Sometimes the piece you were going to leave behind ends up being the perfect piece for your collage.

2. Select your base layer pieces for however many minis you'd like to create and set aside.

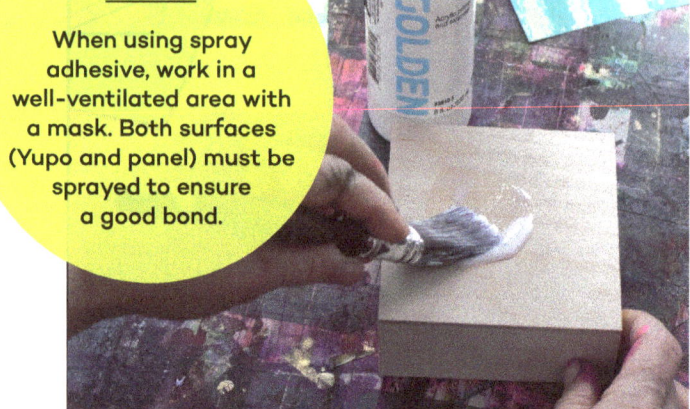

NOTE

When using spray adhesive, work in a well-ventilated area with a mask. Both surfaces (Yupo and panel) must be sprayed to ensure a good bond.

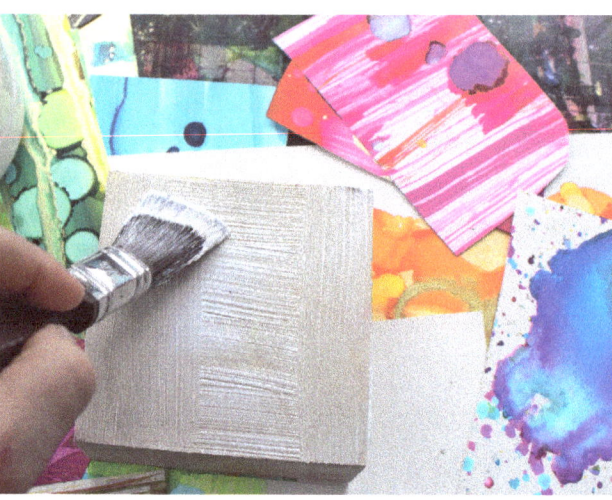

3. First, seal your wood panel. An artist sealer such as Golden GAC 100 works well. Squeeze some onto the panel and brush it across the surface until covered. Once the sealer is dry to the touch, you can move on to the next step.

4. For your base layer adhesive, use a medium body matte or gloss gel, spray adhesive, or YES Paste.

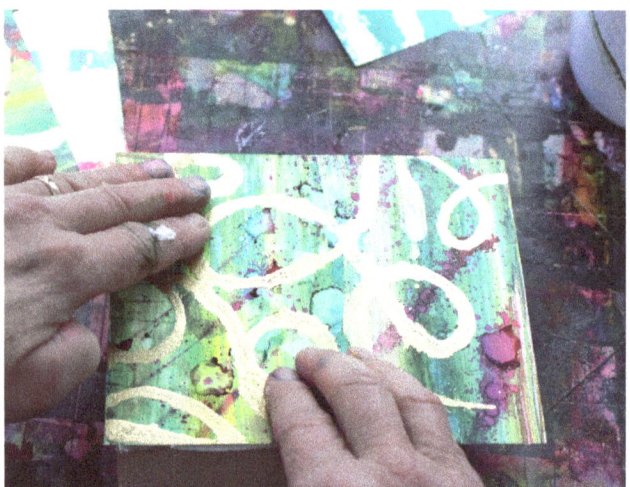

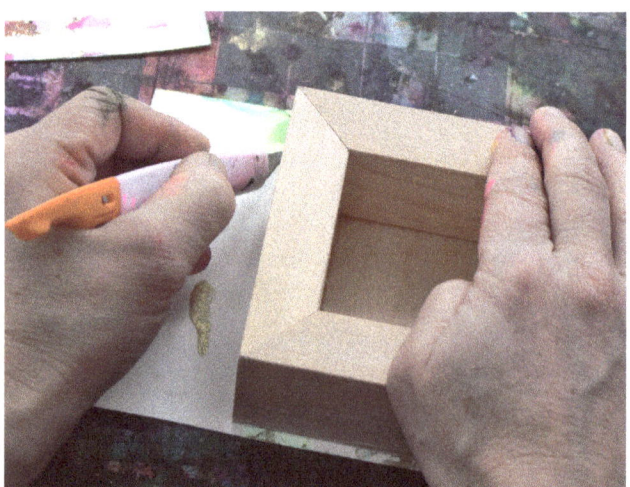

5. Firmly press your piece of Yupo paper onto each wood panel. Rub over the surface with some pressure in a circular motion to ensure good adhesion and to remove any air bubbles.

6. Turn the panel over, hold down in place with one hand to steady it, and trim the excess paper with an X-Acto knife.

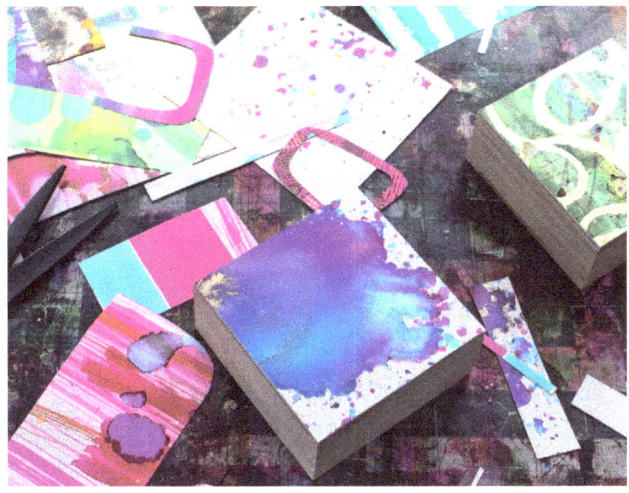
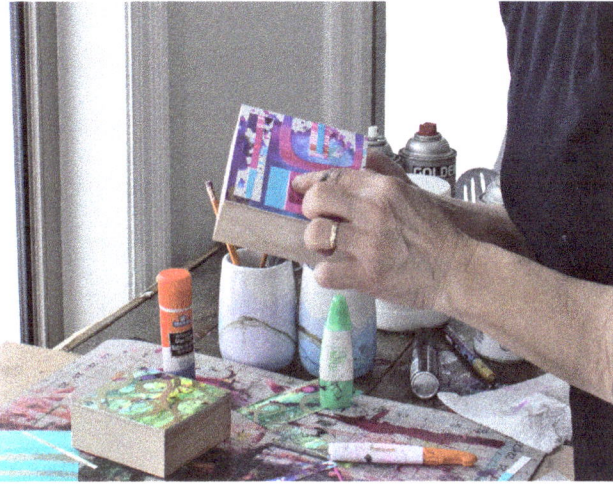

7. With your base layer down, you can begin deciding on which colors and shapes you would like to see on your next layer. (You can leave a panel with just a single layer if you don't want to create a collage.)

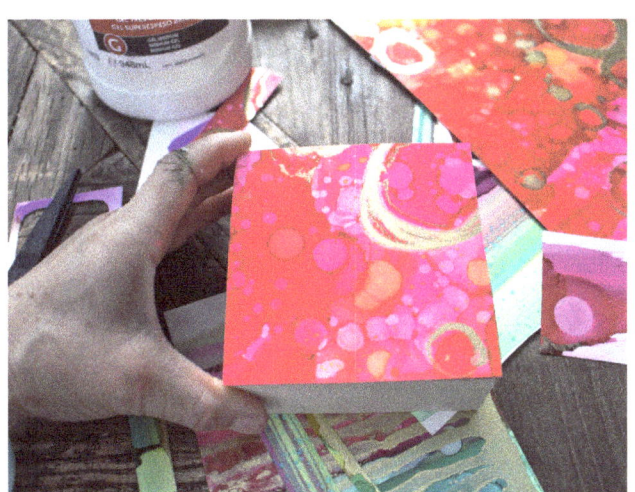
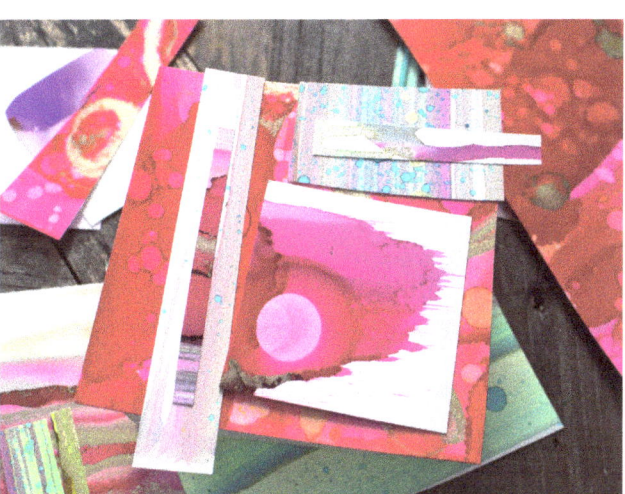
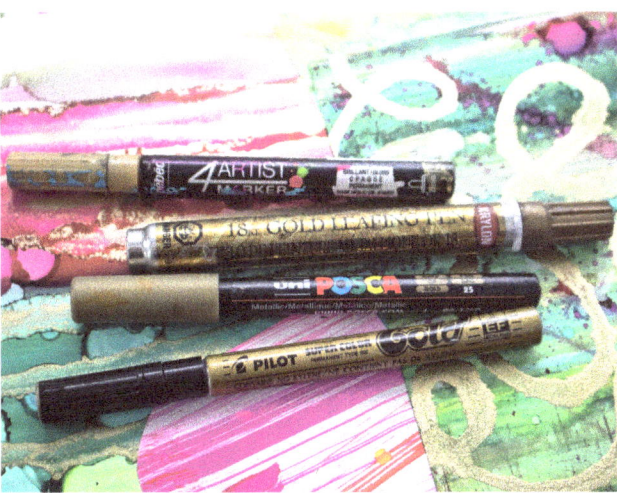
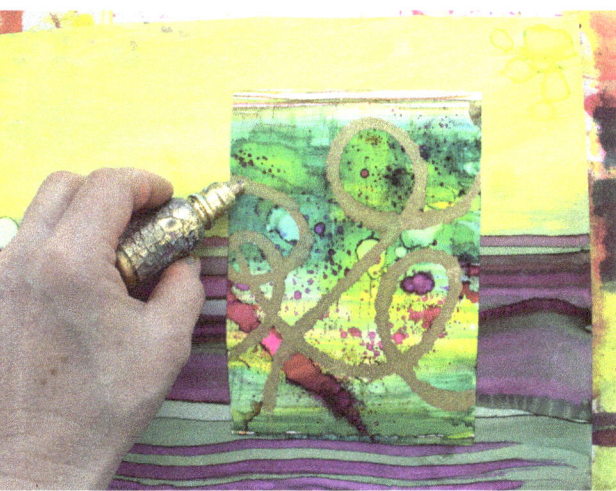

8. Start cutting and arranging some shapes on your panels to get a visual of how they will look. Coordinate colors for an attractive look. Use a color wheel if you're not sure which colors work well together. You may want to add some extra embellishing with gold metallic markers or metallic alcohol ink to give them some extra shine, texture, and appeal.

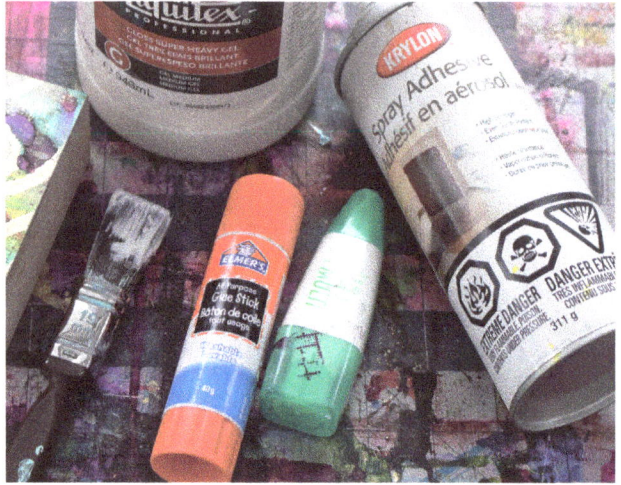

9. Once you've determined your layout, you can begin adhering the shapes onto the base layer. Different adhesives and applicators may be easier to work with depending on the size of the shape you are adhering. You can use glue sticks, thin-tip glue pens, broad glue markers, and more. Experiment with different types to find what works best for you, but be sure to choose a glue that dries clear.

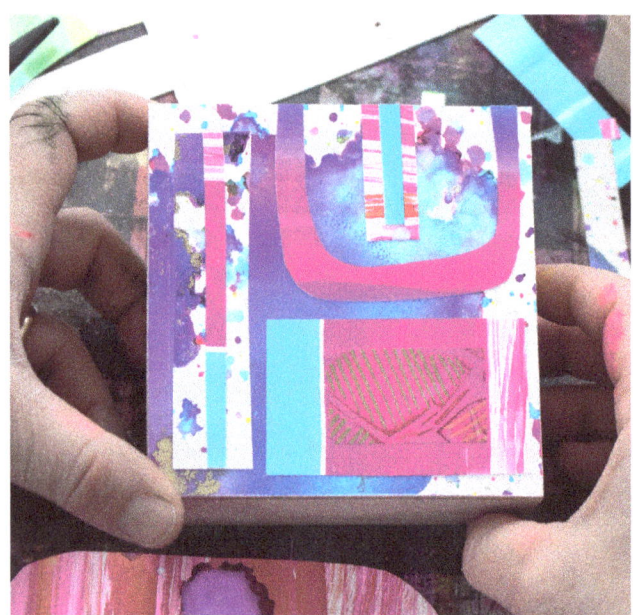

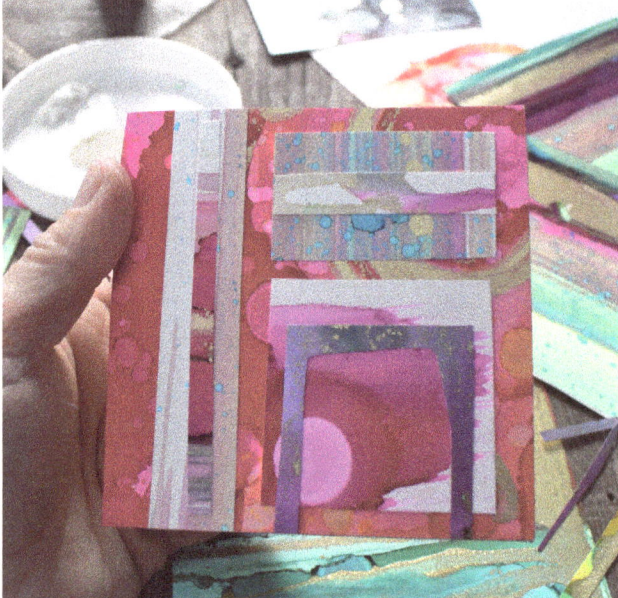

10. Once your pieces are glued down, trim off any pieces that hang over the edge to create a smooth, clean look.

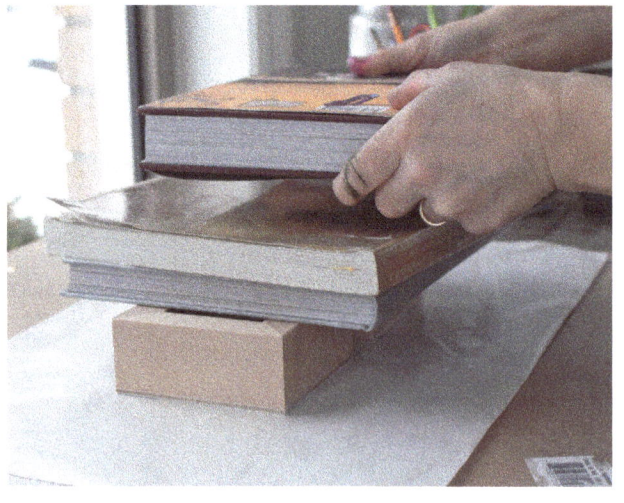

11. To ensure a good bond while drying, place your MOD Minis upside down on a flat surface covered with parchment paper. Weight them down with some heavy books and leave for at least 24 hours before moving on to the next step.

TIP

Take extra care with properly adhering your second layer of cuttings. If a good bond is not achieved, bubbles can get trapped between layers when applying resin.

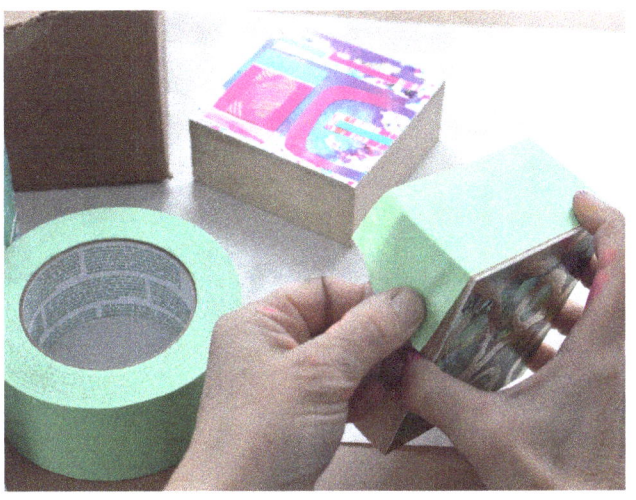
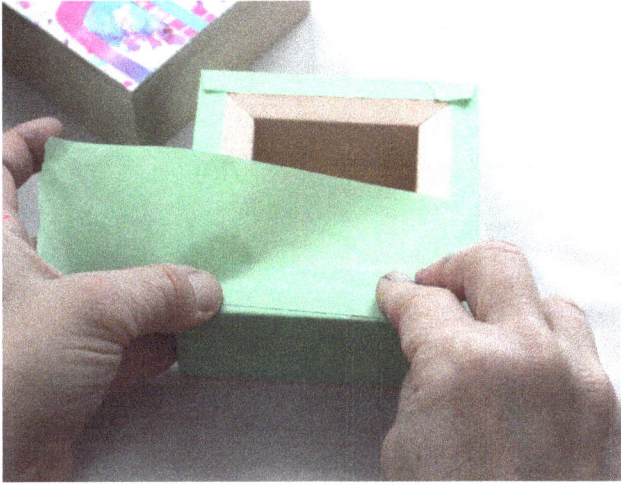

12. After one full day, remove the weights from your panels. Apply painter's tape to the edges just below the paper line and the underside of the panels. Any brand of painter's tape is fine. This will keep the sides free from resin. Smooth the edges with your hands to make sure the tape adheres well to the wood.

TIPS

- Always tape underneath your panel frame to prevent resin drips from curing on the underside of the wood.

- When applying sprays, work in a well-ventilated area (preferably outside) with a mask.

- Place your minis in a large cardboard box or container while spraying to prevent the adhesive from settling onto unwanted surfaces. Spray adhesives are extremely sticky and messy.

13. Before moving on to the resin stage, it's very important to seal and protect your alcohol ink art. While most resins on the market today contain a UVLS (ultraviolet light stabilizer) that protects your art from fading from the sun, you will want to take added measures. Depending on the brand of inks you use, some tend to bleed more than others when a varnish or resin coat is applied. Sealing your art first will prevent this from happening. The Krylon line of sprays are very popular among alcohol ink artists and hold up well on many different substrates. If you can't find Krylon in stores or online, alternative products can be found in the Resource Guide section of this book.

14. Follow the product instructions and shake the can well for at least one minute. Test the nozzle by spraying off to the side first to make sure the product is spraying consistently. Always test products on something else first before applying to your final work. Start with the sealant product spray in a sweeping motion until the art is covered in a fine coat. Allow the product to dry and then apply the UV protectant spray. Leave your minis to dry for several hours before applying the resin.

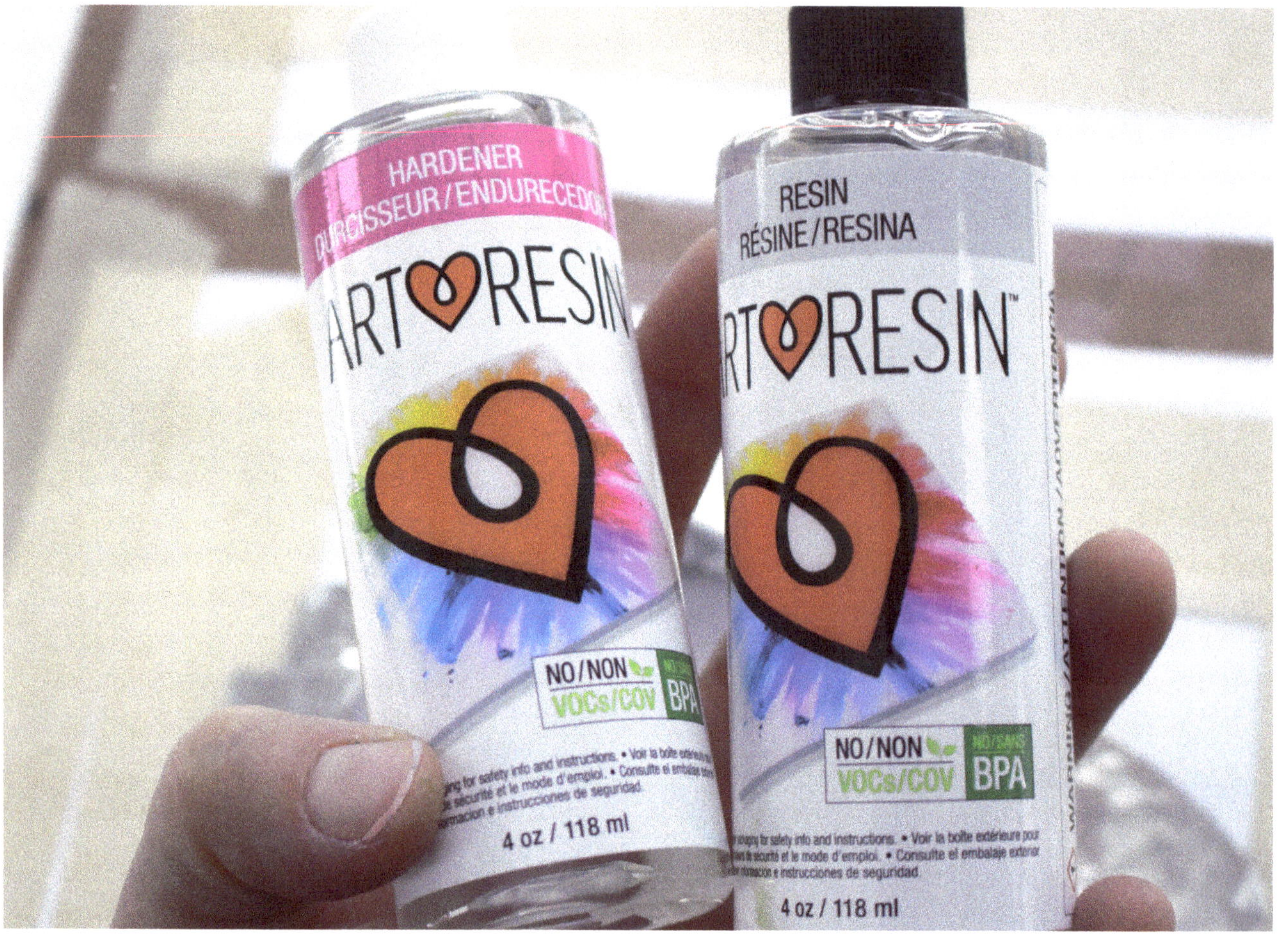

RESIN STAGE

Many people are afraid to try using resin because it appears intimidating, but I can assure you once you get the hang of it, you'll be hooked. Resin not only looks beautiful, but it also adds a protective barrier, lends depth, and brings out the color of your art, especially when used over alcohol ink. When done properly, the results can be stunning!

There are many brands of resins on the market and they are all created for different purposes. There are industrial-purposed resins, casting resins, and resins made for your art. Each brand of resin has its own mixing and measuring instructions and working times. Nearly all brands, however, work in a two-part epoxy system, which means you need to mix together two components—a hardener and the resin itself.

It's best to research and start with sample kits to determine the best resin for your needs. Choose to work with a resin that contains zero VOCs and is low odor. For additional safety, wear a mask.

Resin is very messy and sticky, so it's important to protect the surface areas you're working on and around. You should also wear powder-free nitrile gloves. You don't want to get resin on your hands!

Resin is generally self-leveling. Before you start, use a level to make sure your work area is even. If it is not, you will end up with a cured layer slighty higher or lower on one side.

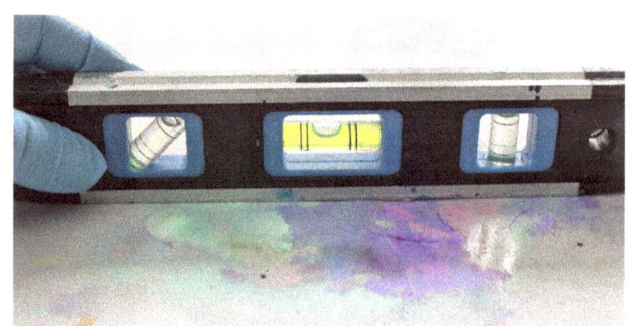

MATERIALS

black garbage bag or plastic tablecloth to protect surface area

level

safety gloves

resin kit

measuring/mixing containers

craft sticks

butane torch

plastic/cardboard box to cover work

sandpaper (optional)

1. Arrange your mini canvases on the surface (you can either place them directly on a non-sticking resin surface like a thick black garbage bag, clear plastic shower curtain/tablecloth, high-quality parchment, or freezer paper), or set your minis on containers such as mini paper cups. Make sure your work area is level.

2. Wearing safety gloves, measure and mix your resin according to the manufacturer's instructions (resins are either measured by volume or weight).

3. Mix your resin slowly and consistently for at least 3 minutes or according to the instructions, scraping the sides and bottom of the container too. The slower you mix, the better, as it will prevent excessive bubbles from forming. Try to work in a room set at 72°F (22°C), which is the optimal temperature for curing. Fluctuations in temperatures during the curing time (Art Resin fully cures in 24 hours) can cause dimples in your finished layer. A consistent temperature is key to getting good results.

TIPS

- Check out the Resin FAQ/Troubleshooting section for frequently asked questions, tips, and troubleshooting.

- If your surface is uneven, use mini bathroom paper cups or small pieces of cardboard and add them incrementally on the corner you need to raise or adjust.

- Art Resin brand is mixed by volume and therefore measured in equal parts. This brand has a longer working time (45 minutes) and has a user-friendly online usage calculator to determine the amount of resin you need for a project. Find more at www.artresin.com.

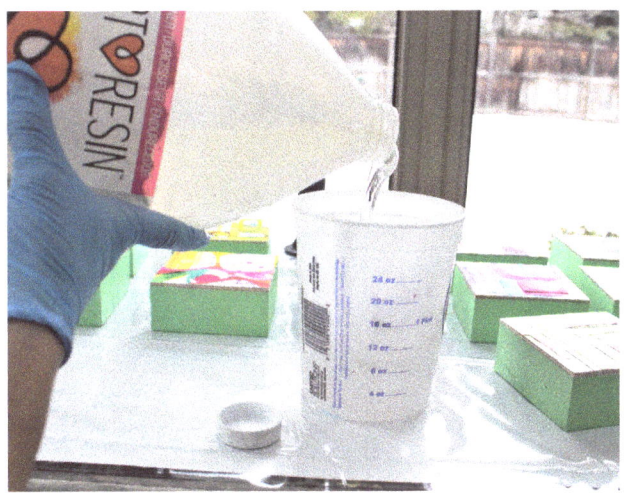

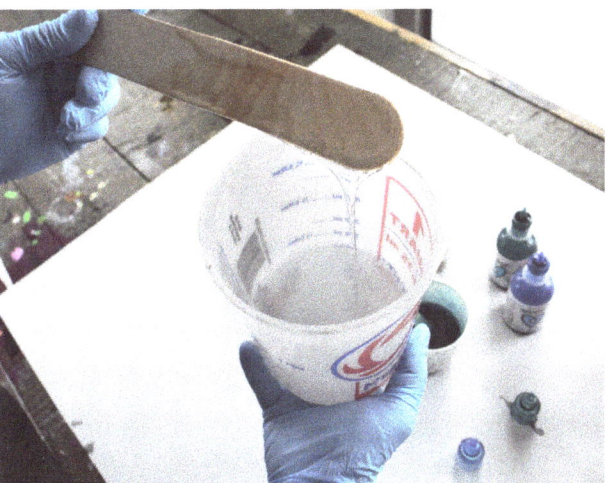

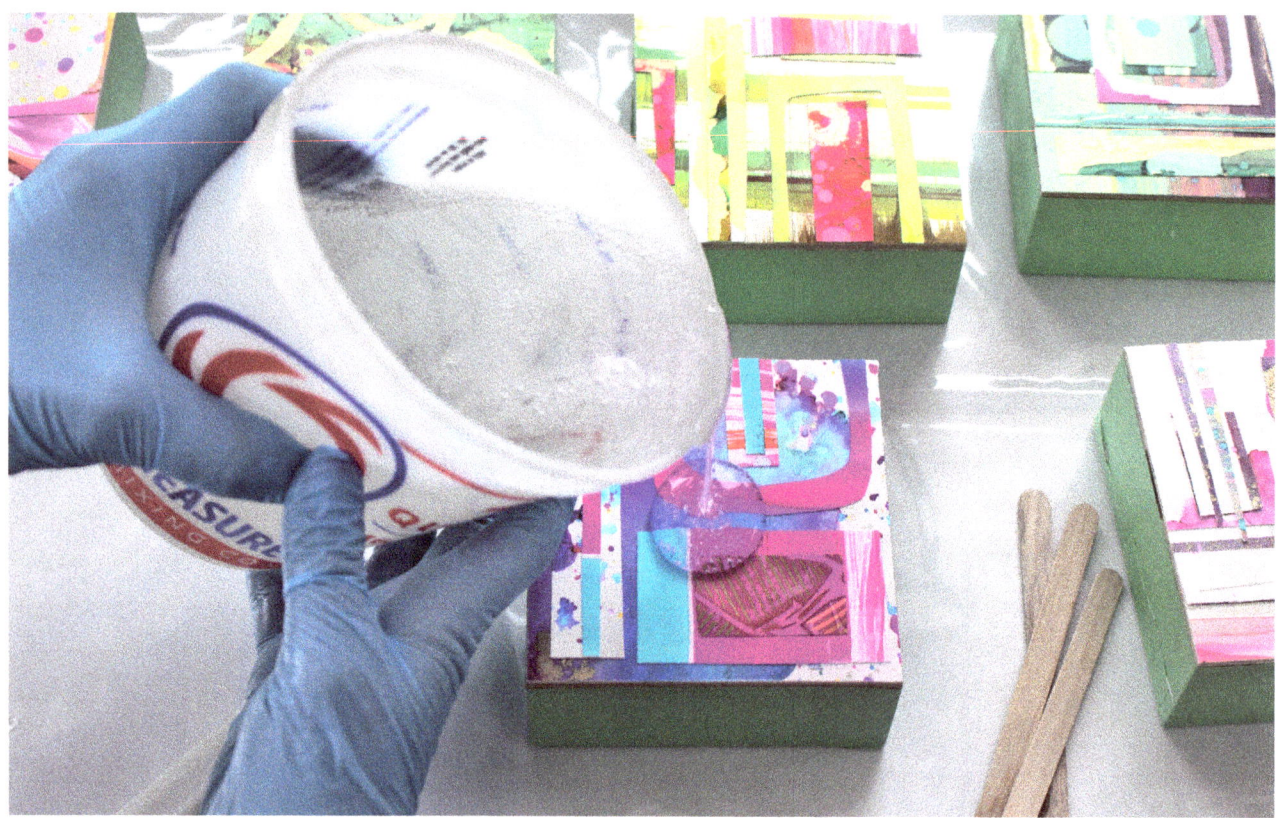

4. Begin slowly pouring some resin onto each panel (don't pour from too high above the surface) until you distribute the resin equally onto each mini.

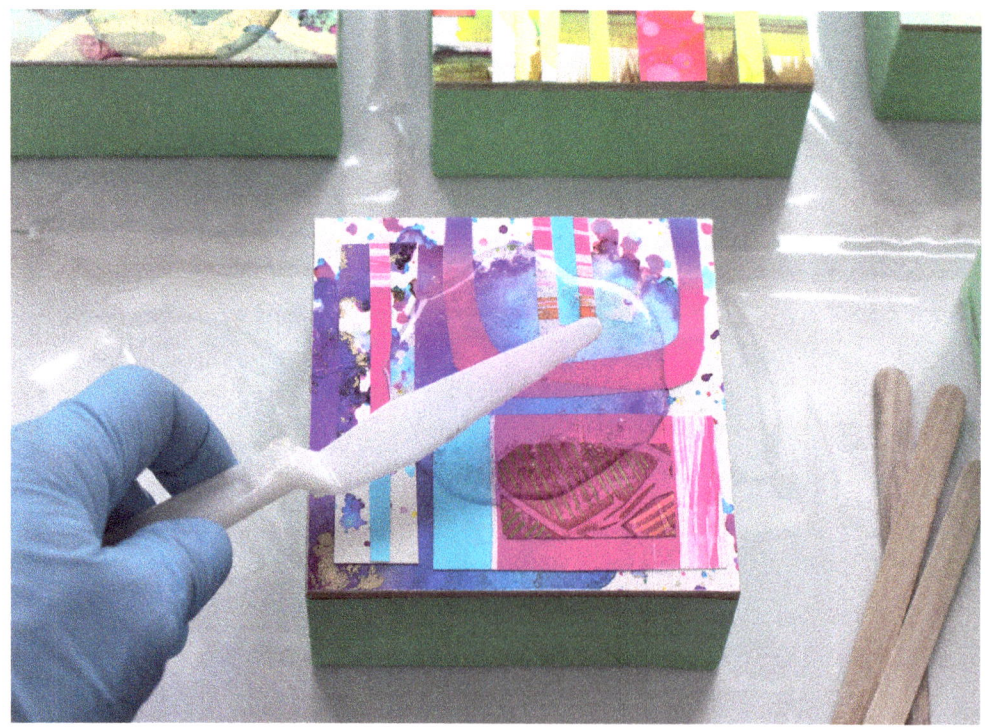

5. With a craft stick, plastic palette knife, or silicone tool, begin moving and spreading the resin across each panel (like icing a cake), covering the surface evenly and paying attention to the edges and corners.

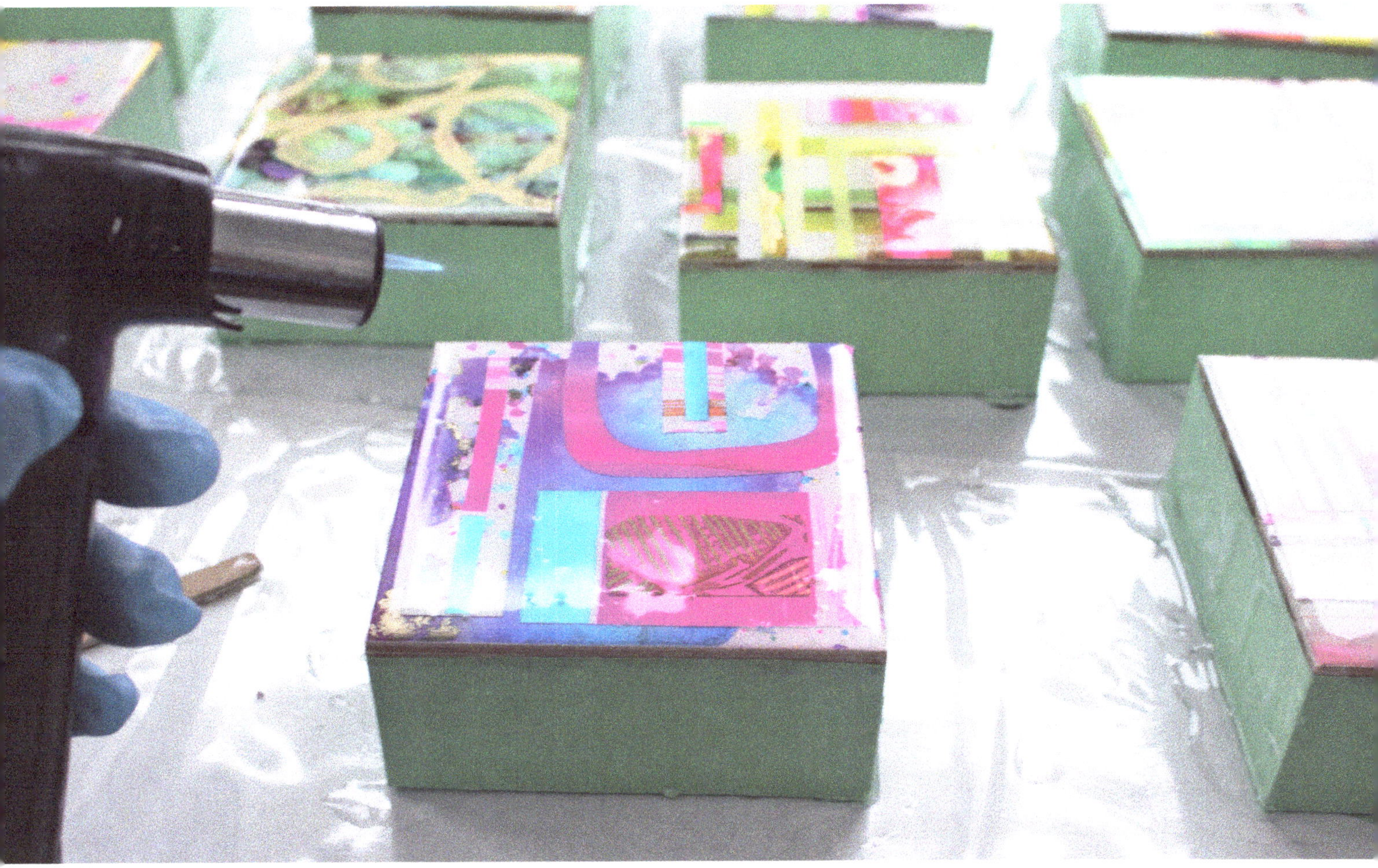

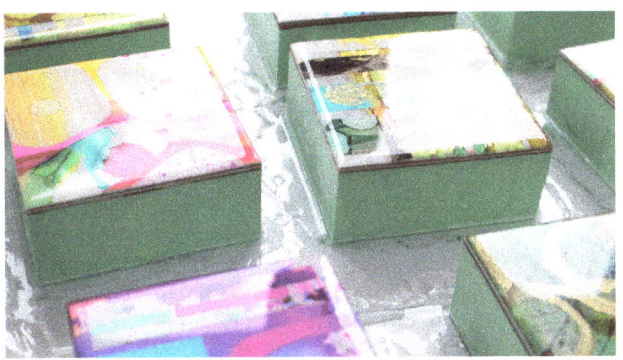

6. When all of your MOD Minis are covered evenly with resin, use a small butane torch to eliminate any bubbles on the surface. This is the best and most effective method to get rid of bubbles. You need only to kiss the surface, moving the torch around quickly without hesitating too long in one area. If you leave the torch on one area for too long it will cause dents and marks in your cured resin. Continue to torch over each panel until the minis appear smooth and glass-like.

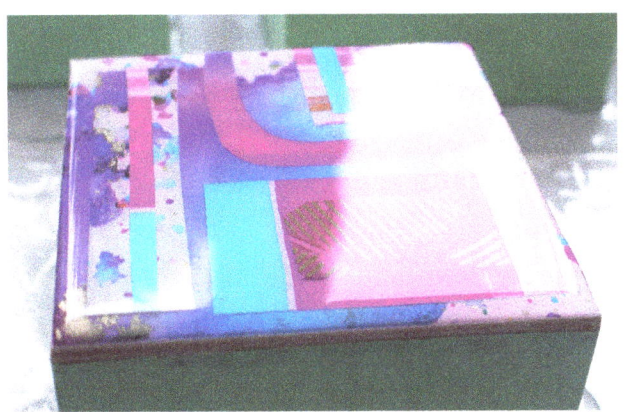

TIP
—
Double-check closely for dust or hair that may have settled into the resin and remove with a toothpick.

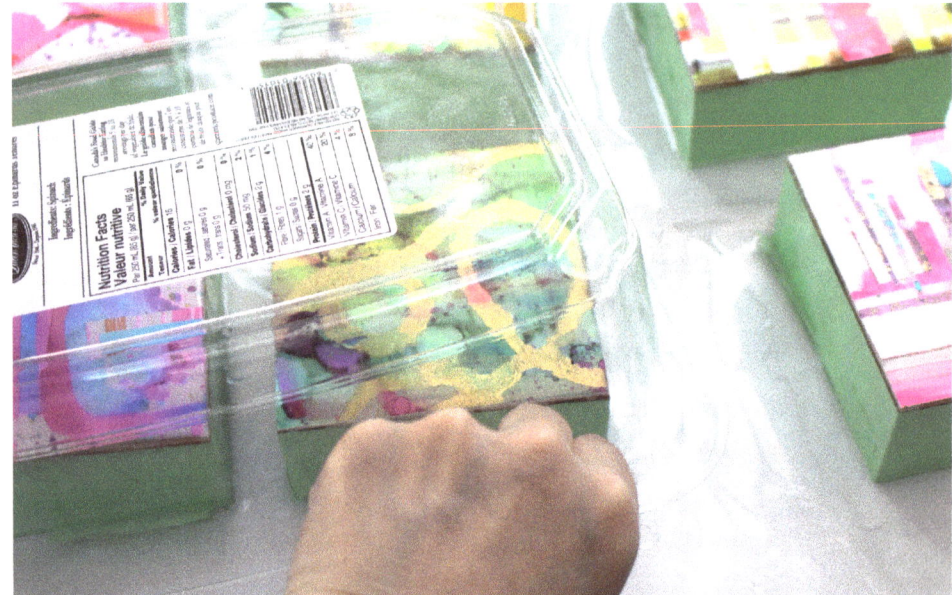

7. When you are satisfied with how each panel looks, cover them with a cardboard box or plastic container to prevent dust particles from landing in the art while the resin cures. Empty produce/salad containers work great for this project!

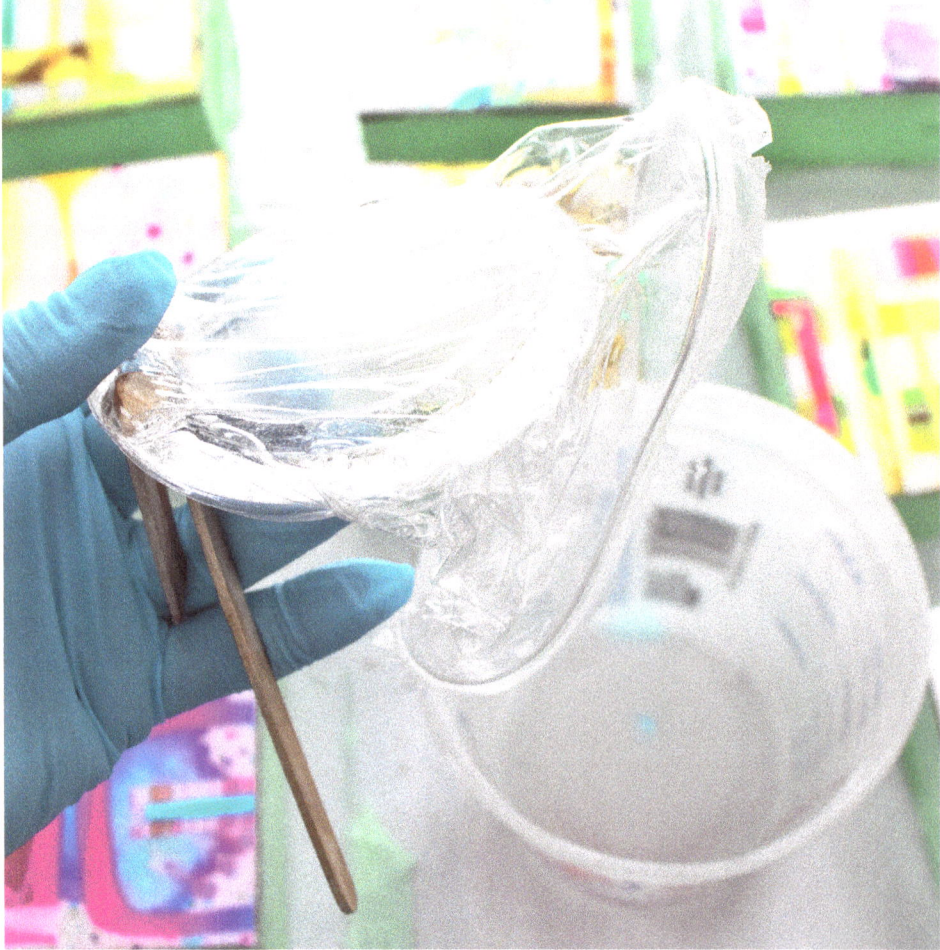

TIPS

- Using silicone tools with resin makes for easy cleanup as cured resin peels right off, creating less waste and allowing the tools to be reused.

- Use a plastic mixing container for your resin. When you've used all the resin in the container, leave a large craft stick in it. The following day, pull out the craft stick, bringing the dried resin film out with it. You now have a clean and reusable container.

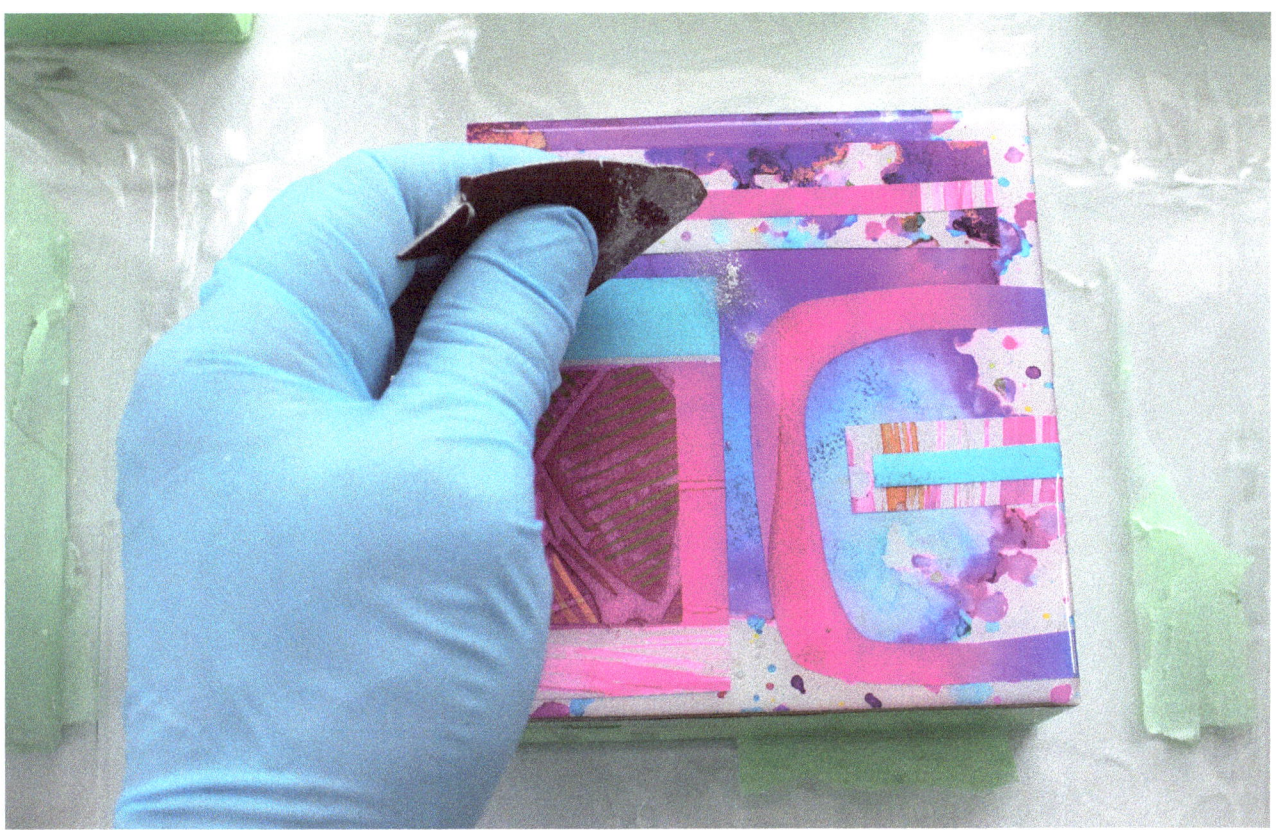

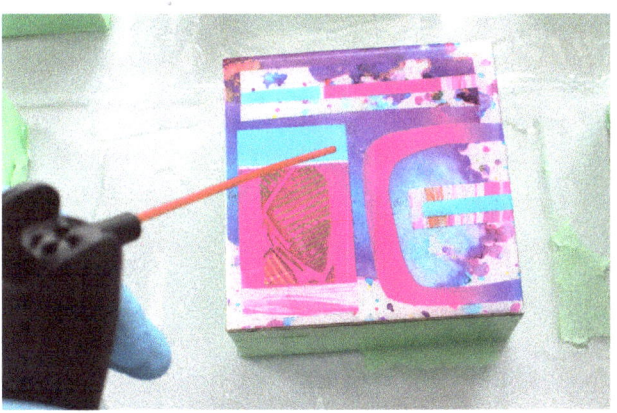

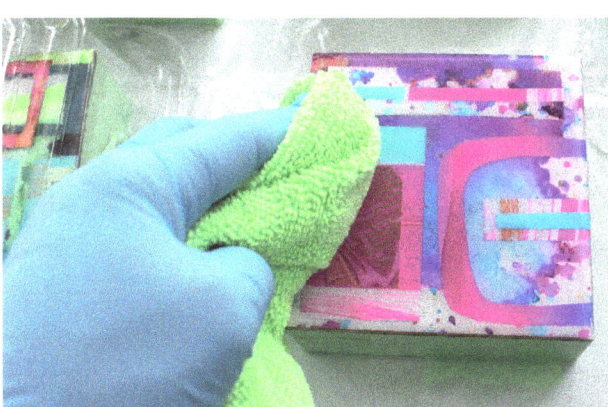

8. Allow the resin to cure for at least 12 hours before checking it and removing any tape. While Art Resin fully cures in 24 hours, it becomes harder to remove the tape at that point.

If your surface did not cure perfectly and bubbles still came to the surface, use a medium-grit sandpaper to smooth it over, wipe with a lint-free cloth, or blow the dust off with a can of air. Then repeat the same steps with a second coat of resin. Sanding over a resin layer will NOT damage it. The sanding marks will not show and you can create a nice doming effect on the edges with a second coat.

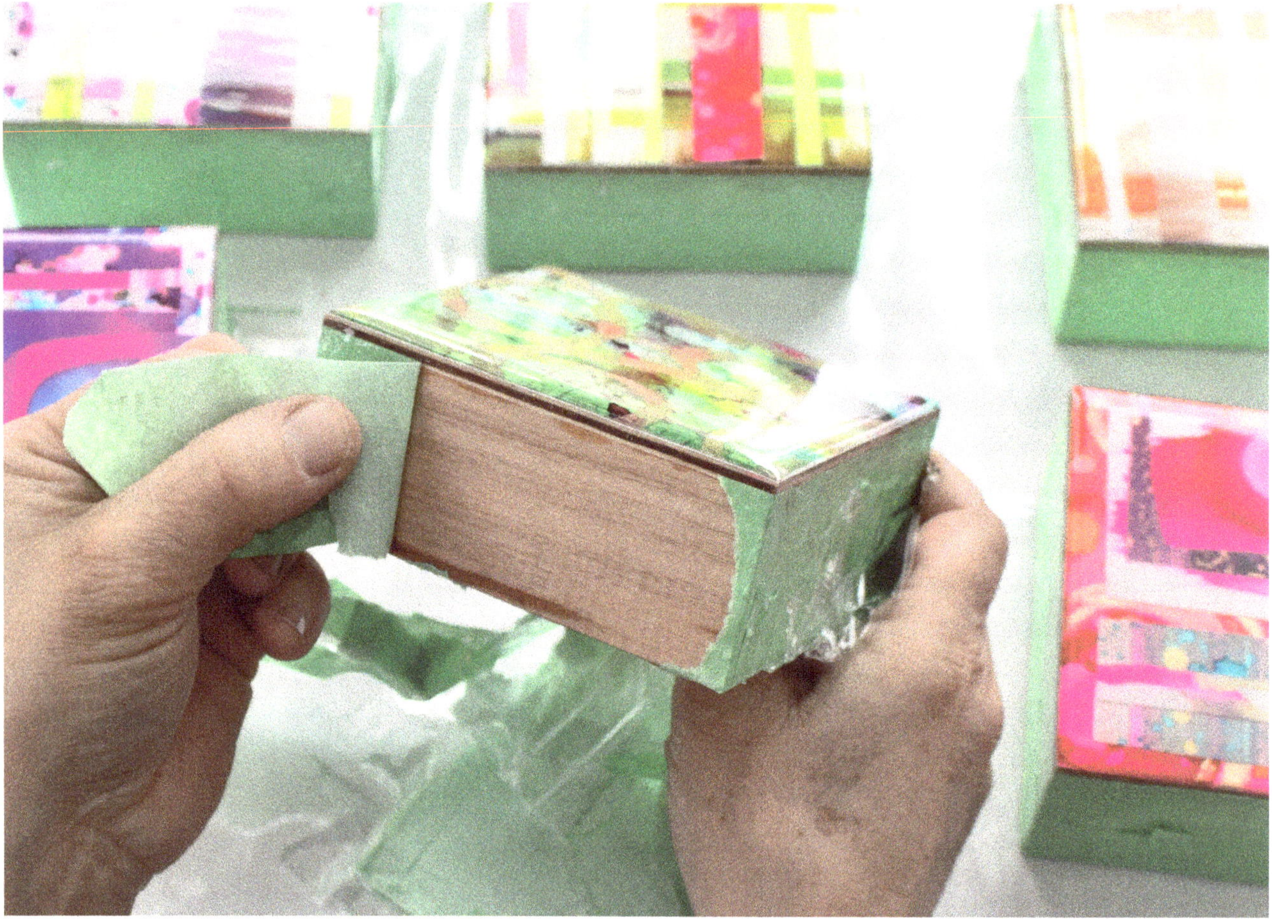

9. Carefully pull your mini art panel off your surface and begin peeling away the tape. The extra work you put in covering the underside with tape will now pay off because you won't have to deal with resin drops that may have cured to the wood.

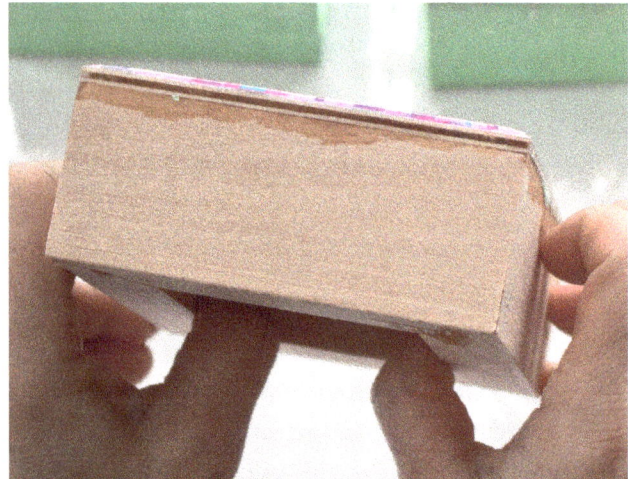

10. After the tape has been removed, you may still notice some resin that managed to seep under the tape. This is sometimes unavoidable, and a few extra steps need to be taken to clean up the sides.

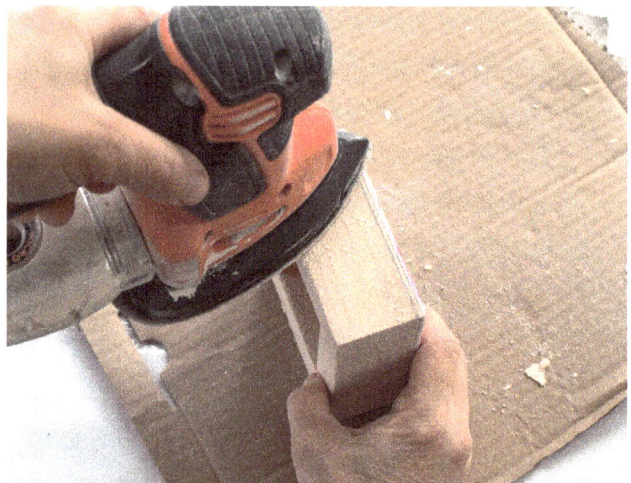

11. You can hand sand using some medium-grit sandpaper, but the best way to produce clean, smooth sides is with a mini electric sander. These are inexpensive and can save you a lot of time and effort.

Voilà! Your MOD minis are complete. Stand back and admire your work. Hang them on your wall with some Velcro strips or set them on your shelf. Either way, they will brighten your day and make awesome gifts!

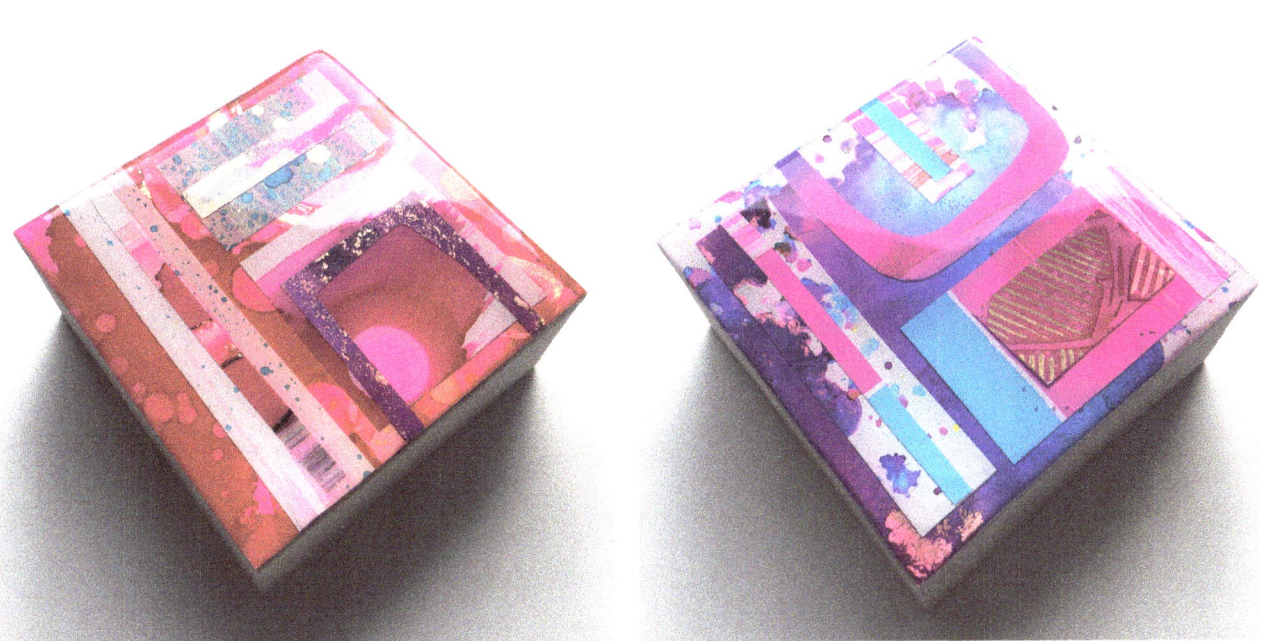

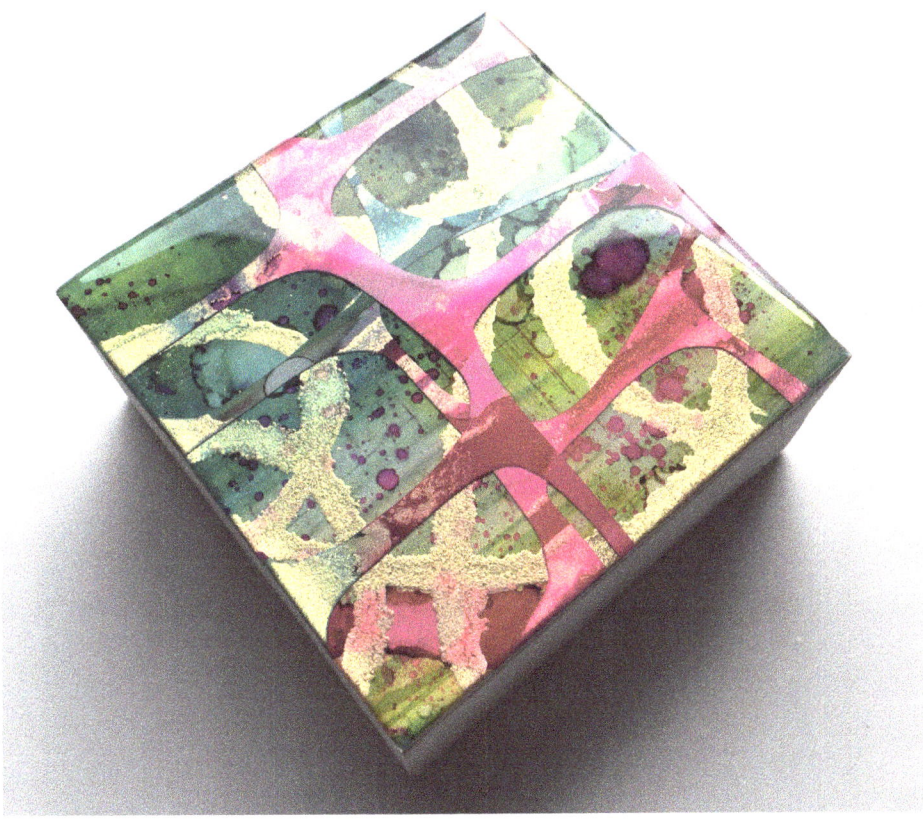

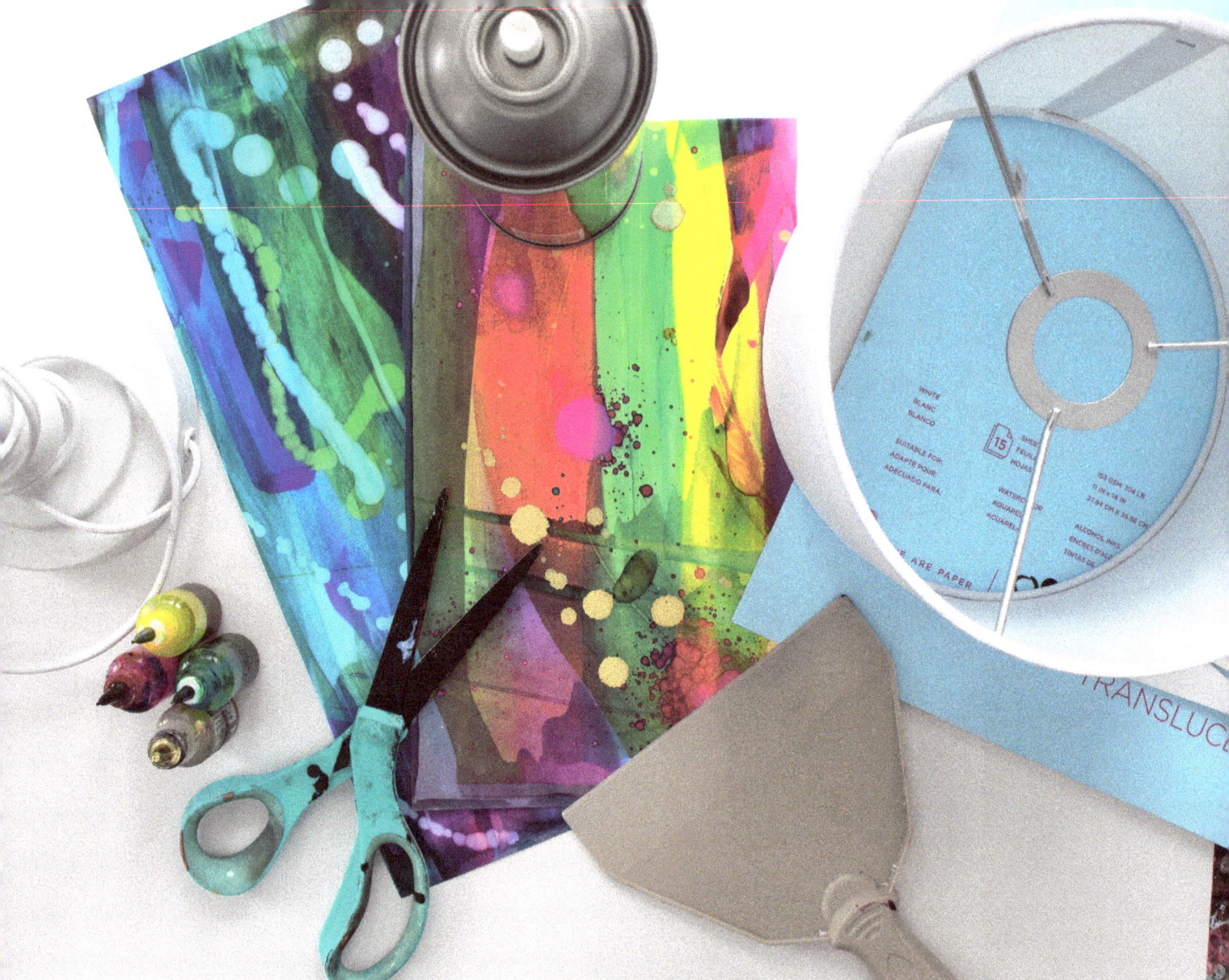

PROJECT 4

ALCOHOL INK DOUBLE-SIDED LAMPSHADE

Custom-created alcohol ink lampshades are vibrant, colorful, and easy to make. They require little effort or skill and the results are stunning. The best part is that you can create a 2-in-1 shade. If you get bored with one design, you can turn it around to reveal another on the other side. The inside of a lampshade is made with a plastic lining and a fabric or paper on the outside. You will be using a plastic composite translucent film to create your design, which you will adhere to the outside of a plain, store-bought shade.

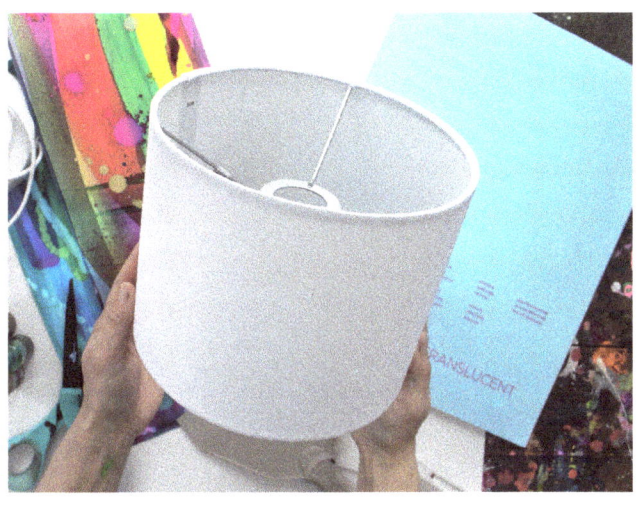
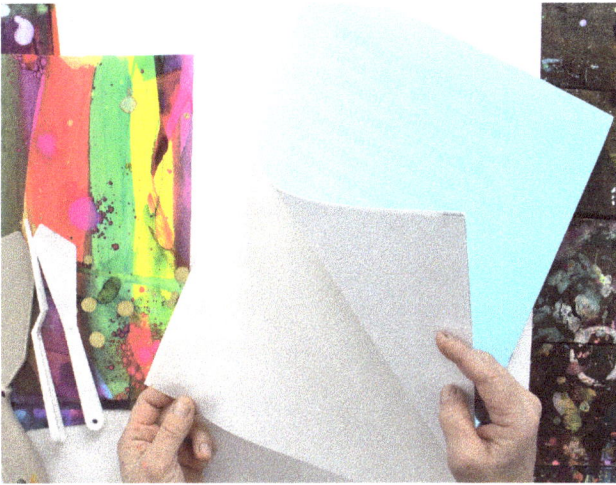

MATERIALS

translucent film, such as Yupo Translucent or Dura-Lar

small plain white lampshade (see Resource Guide for more information)

alcohol inks

plastic scraper

high-tack spray adhesive, such as 3M super 77

scissors

lamp base

1. Begin by making sure the width of your film is wide enough to cover your entire shade. (Used here: Yupo Translucent 11" x 14" [28 x 35.5 cm] pad sheets.) Two 11" x 14" (28 x 35.5 cm) sheets will create a double-sided design that will cover the 25" (63.5 cm) circumference of the shade (the length of the shade if it were opened up and straightened out to a line segment).

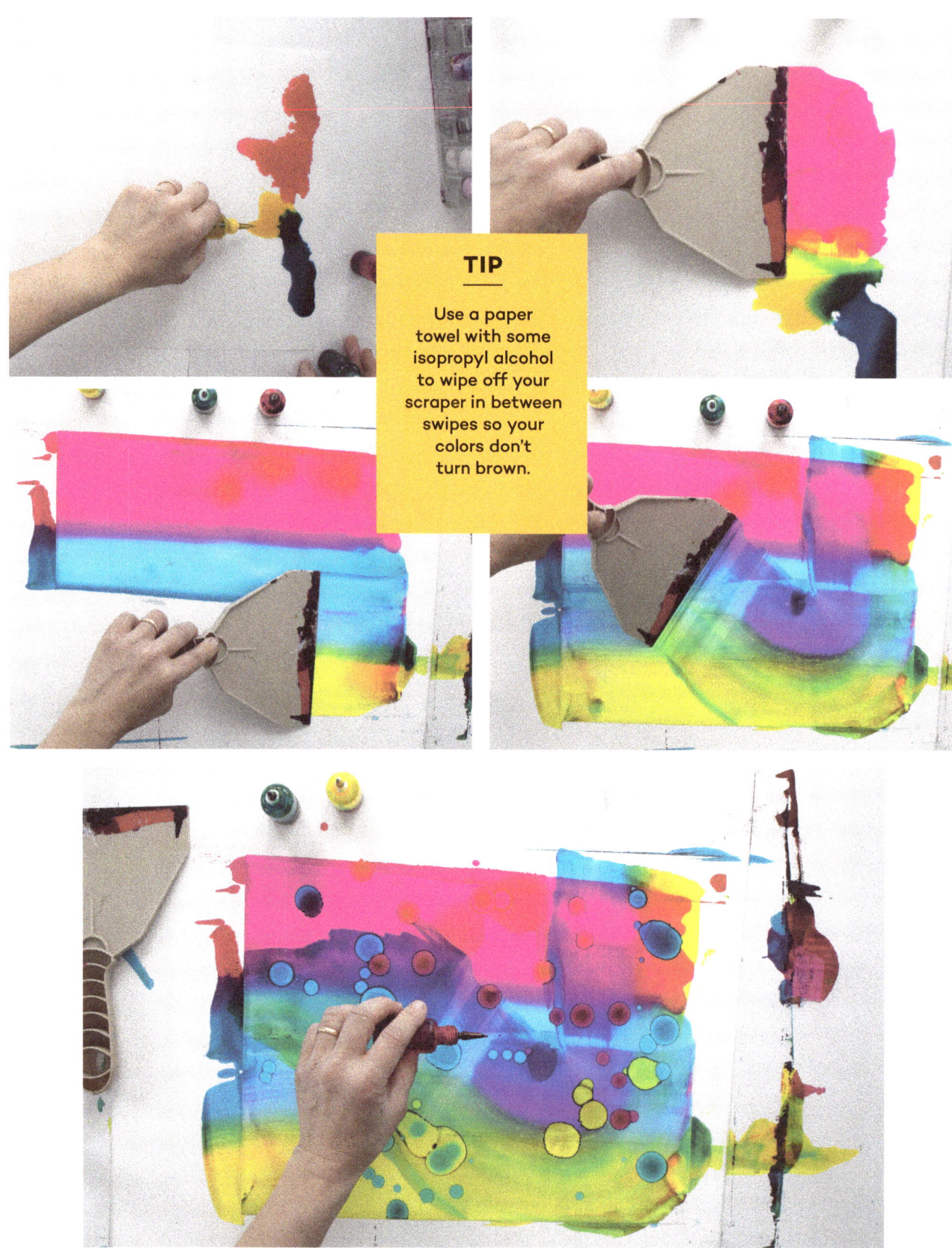

TIP

Use a paper towel with some isopropyl alcohol to wipe off your scraper in between swipes so your colors don't turn brown.

2. Choose some alcohol ink colors (used here: yellow/pink/blue) and begin by adding some drops on one end of the translucent film. With your plastic scraper, drag the inks across the surface to the other side. It doesn't have to perfect. Be random with your design and drops of ink.

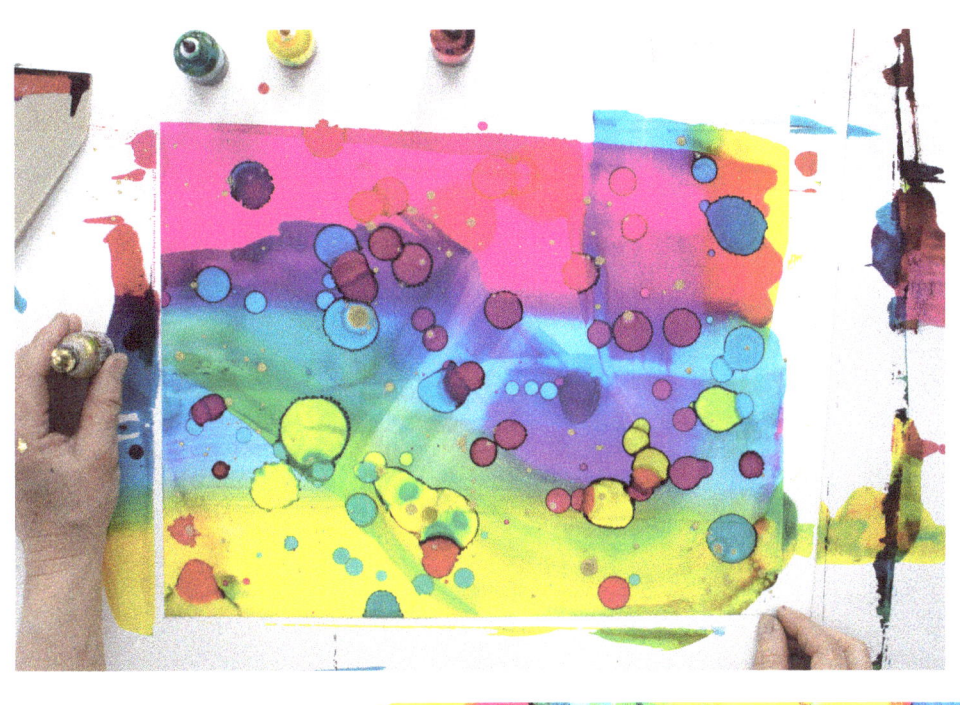

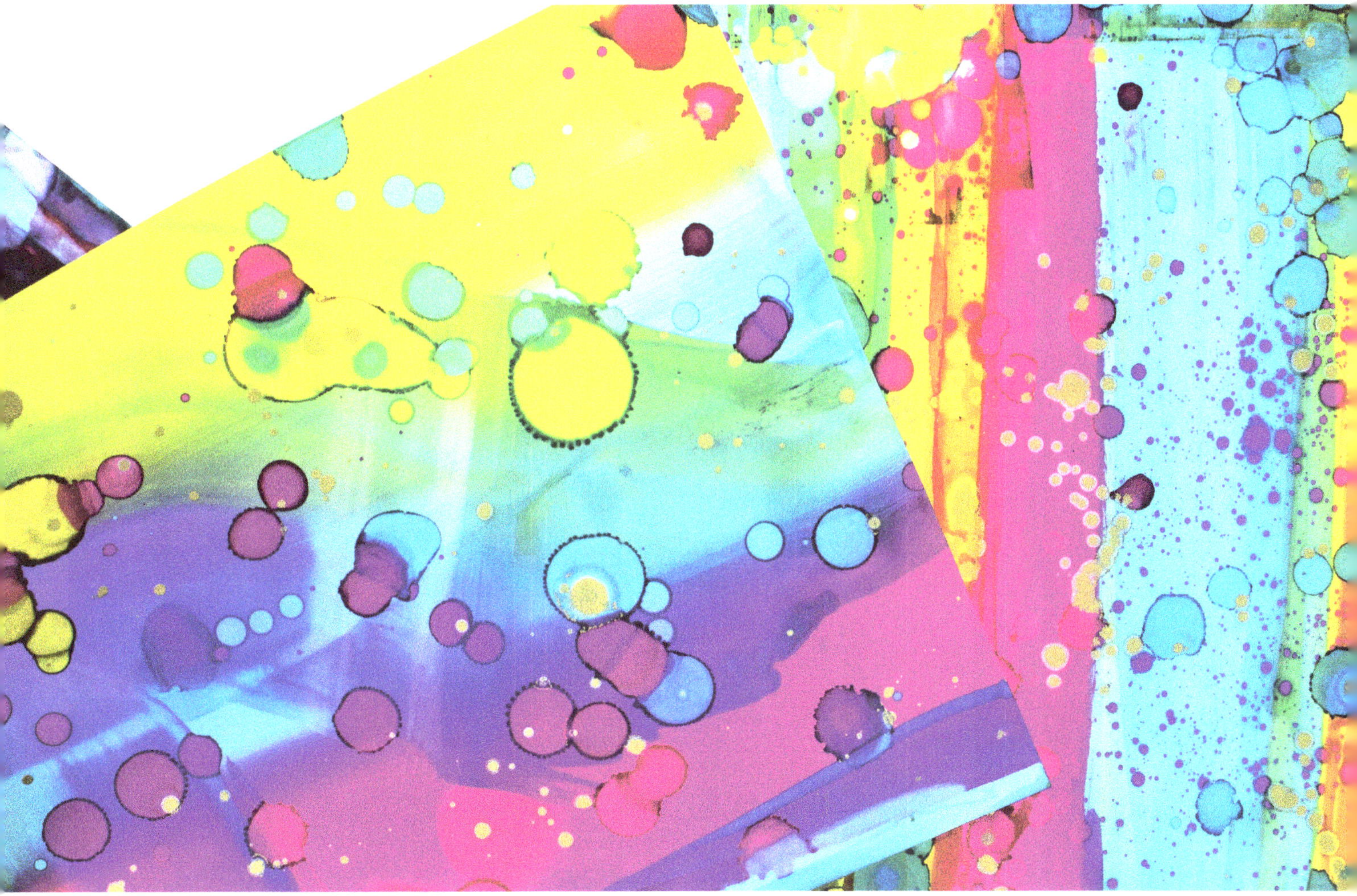

3. When you are happy with your design, allow it to dry for a minute or two. Continue to make several others with different colors. When you have finished, select two of your favorite designs to use on your shade.

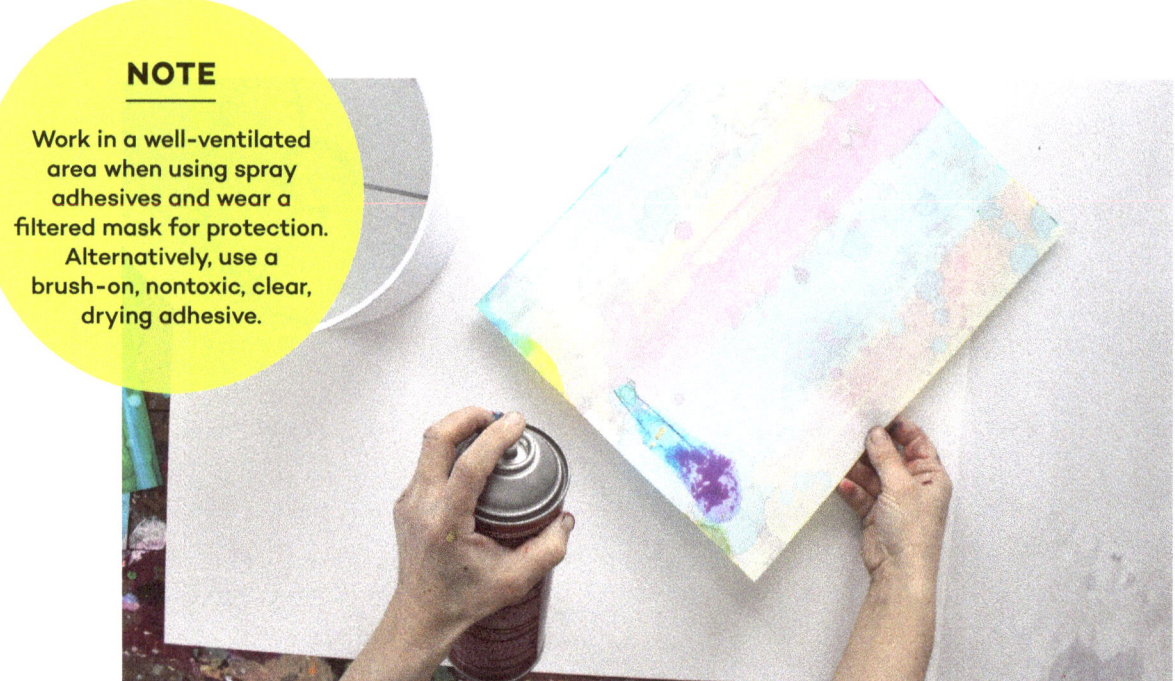

NOTE

Work in a well-ventilated area when using spray adhesives and wear a filtered mask for protection. Alternatively, use a brush-on, nontoxic, clear, drying adhesive.

4. Next, attach your film to the outside of the shade. It is important to ensure your work area is protected well, as the spray is very sticky as it dries. You will want to spray *both* the film *and* the shade for a solid, permanent bond. (Used here: 3M Super 77.)

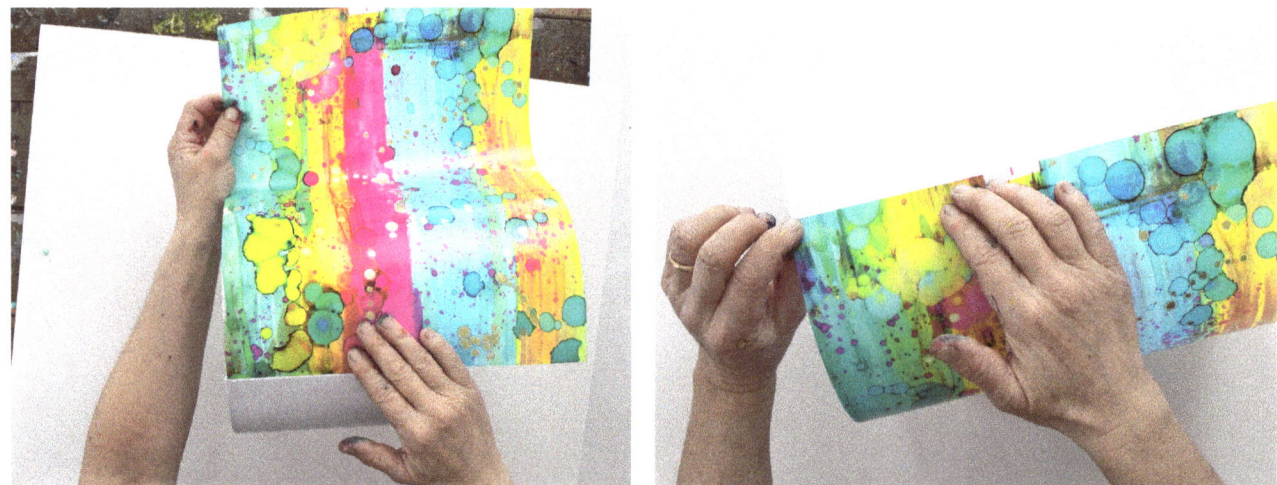

5. Align the sheet with the edge of the shade, where the seam meets. The edge of the shade will be your guide as you slowly turn and secure the design in place. It will also help in ensuring the sheet is straight. Smooth out the film with your hands as you turn the shade so it is nice and flat, without bubbles or air pockets. Continue turning around the drum until your entire sheet is adhered.

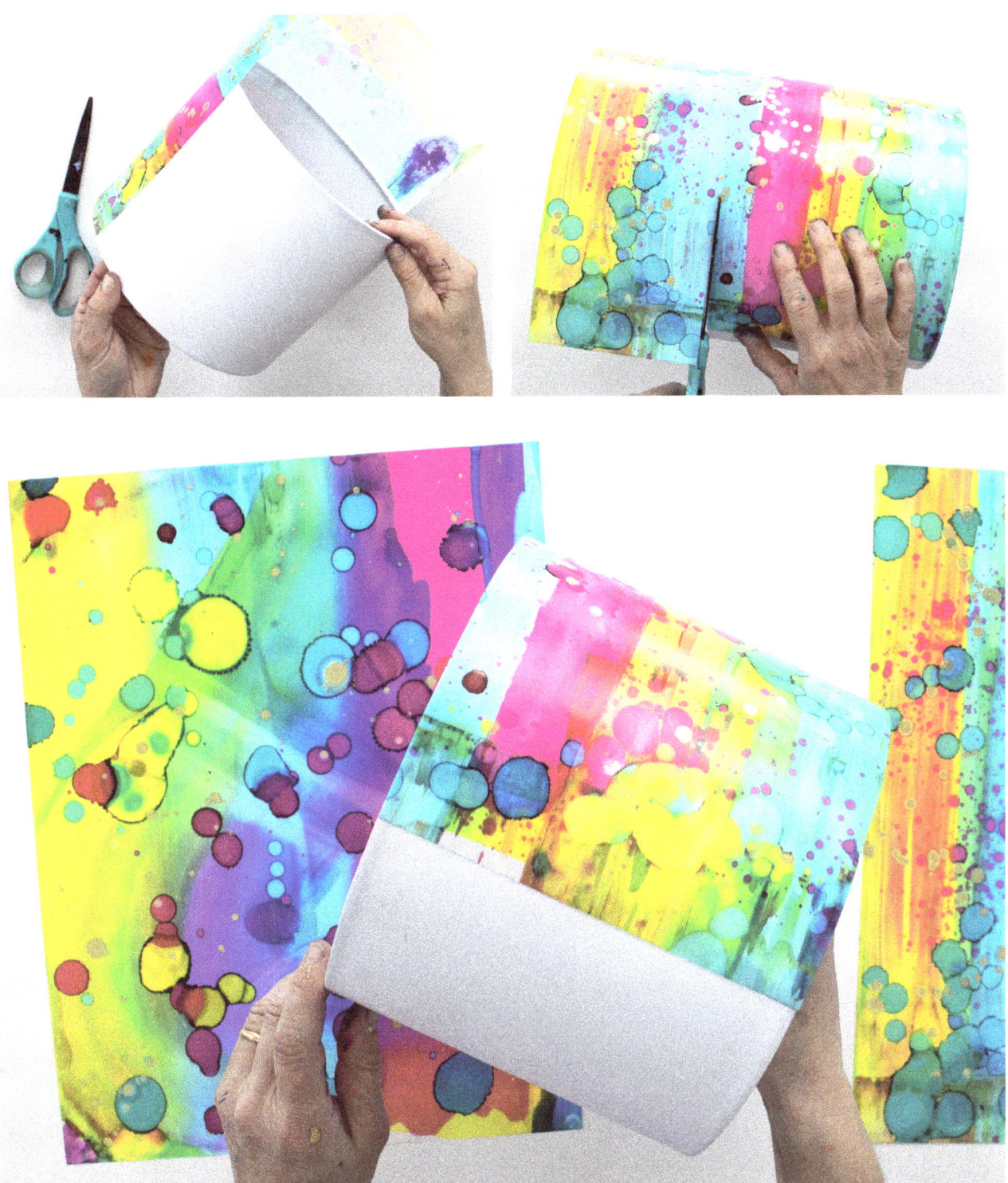

6. You will now use the other side of the shades edge as a guide to cut and trim the excess film. Trim with scissors.

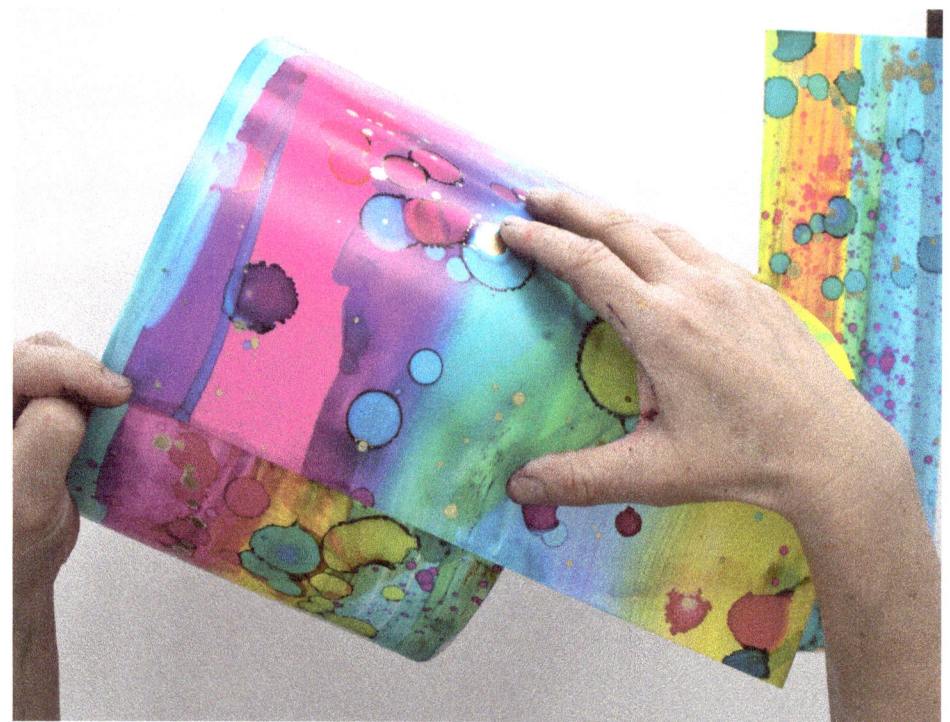

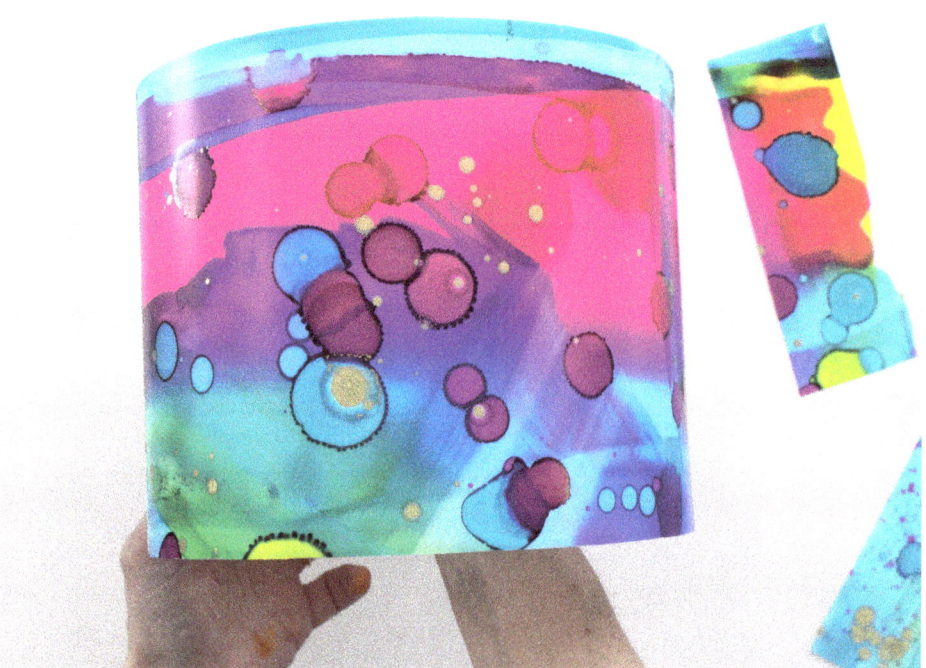

7. You are now ready to repeat the process with your second sheet. Slightly overlap the first sheet, turn, and secure your second piece around the drum until you meet the seam on the other side. Trim and cut up the side of the shade's seam line, so it is nice and straight. Cut the rest of the overhang film around the shade.

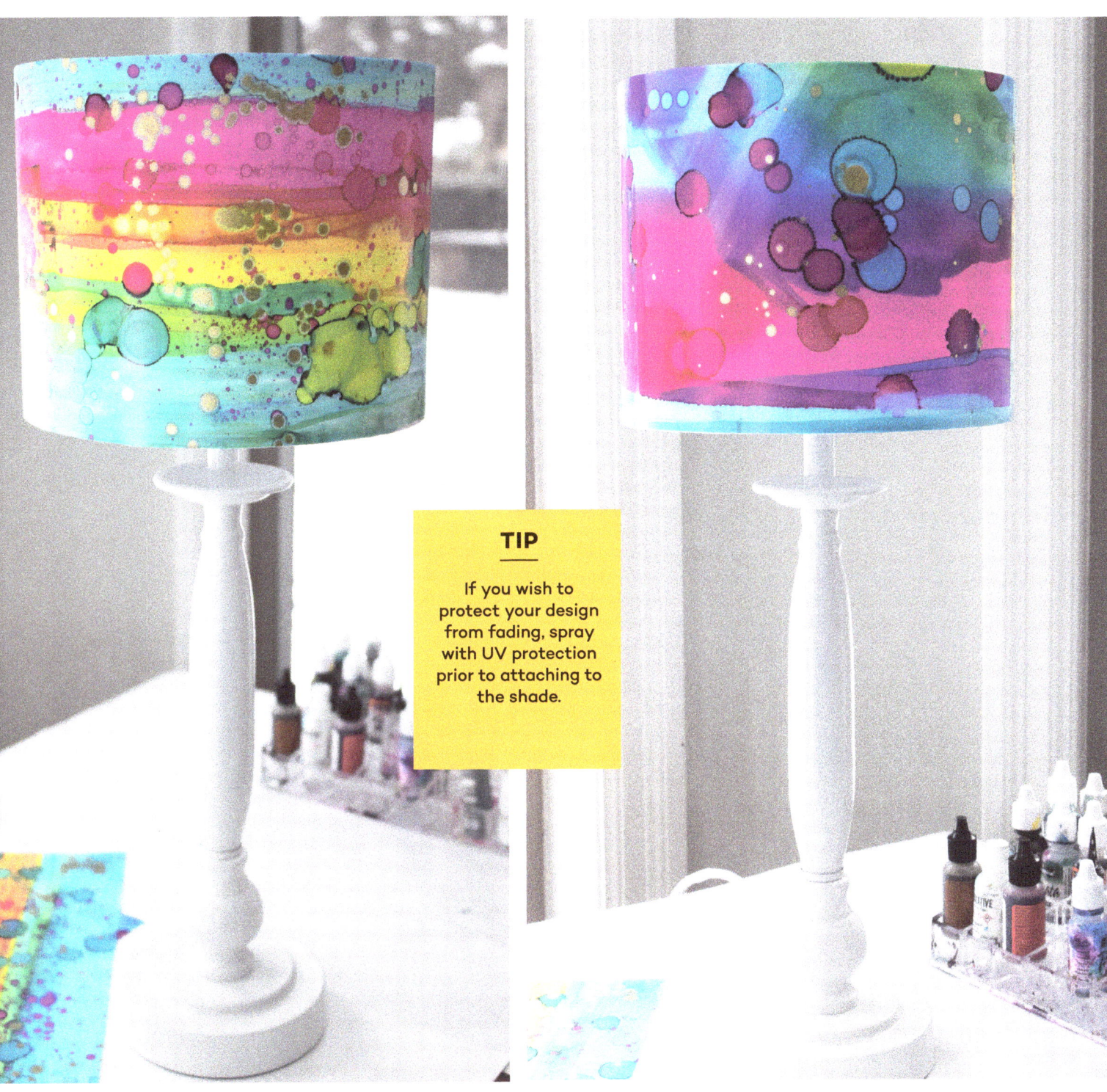

TIP
___

If you wish to protect your design from fading, spray with UV protection prior to attaching to the shade.

Your shade is complete! Attach it to your lamp base and turn to view your double design shade. You do not need to protect your design unless it is sitting near a window. Light bulbs do not emit ultraviolet light!

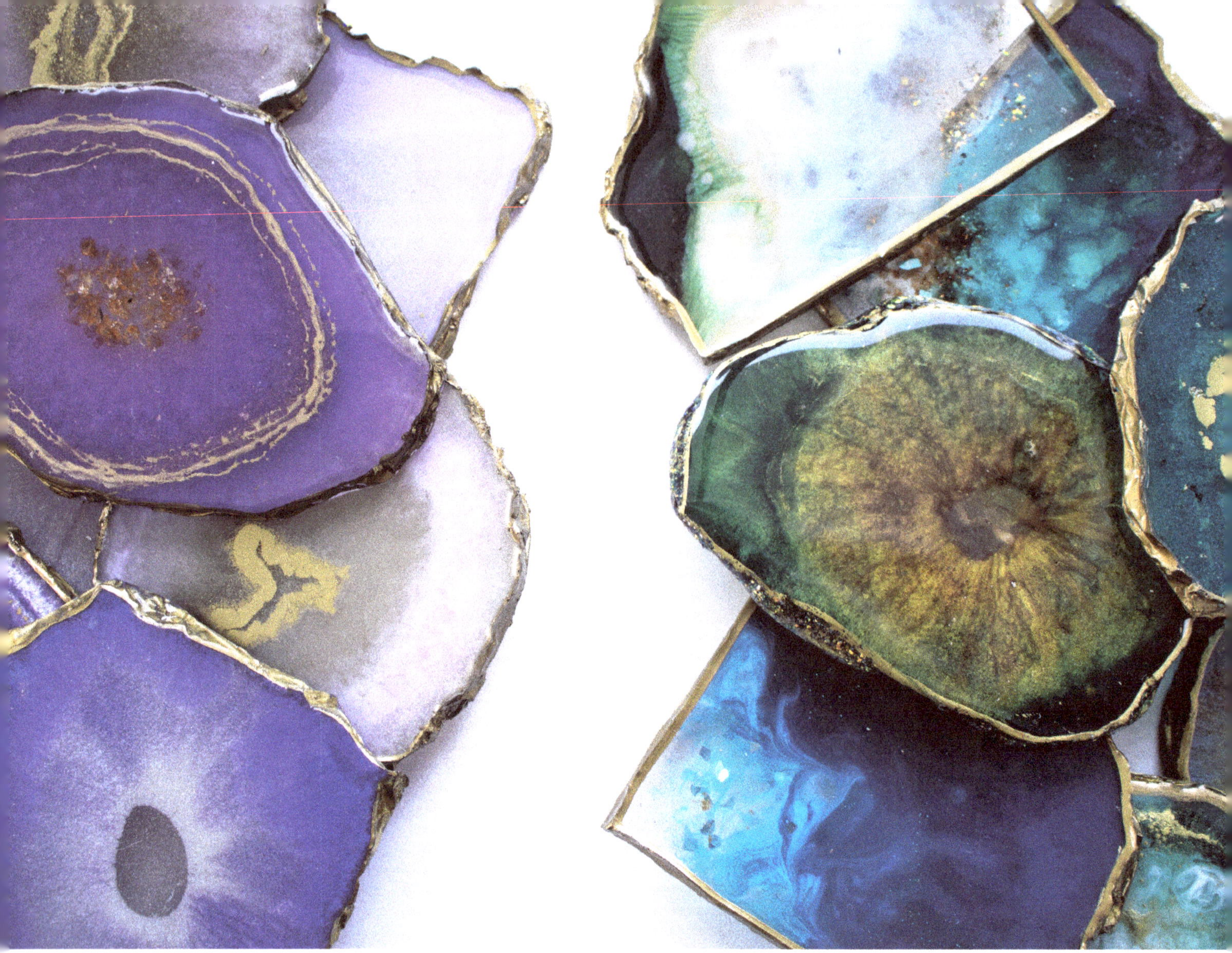

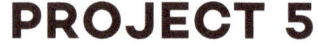

PROJECT 5

GEODE/AGATE SLICE COASTERS

While resin petri molds have exploded onto the art scene, so have geodes and agate slices. Agates are minerals with colorful depth, and when cut into slices, reveal rings and glass-like markings. It has become a trend to create agate-style coasters using resin, resin tints, mica powders, glitters, and glass shards to replicate these gems. There are many silicone molds you can buy to create agates as well as handmade alternative options, such as using silicone and a caulking gun to make a mold shape. But to make truly unique and realistic agates and geodes, use aluminum tape to produce the jagged-style edges (plus it doesn't smell of silicone!). The shapes you can create are endless. From square, to round, to oval, the design is up to you!

MATERIALS

aluminum siding tape

white paper

tape

pencil

scissors

heavy-duty clear plastic shower curtain or tablecloth

glue gun

resin kit

resin tints or mica powders

mixing and measuring containers

craft sticks

butane torch or heat gun

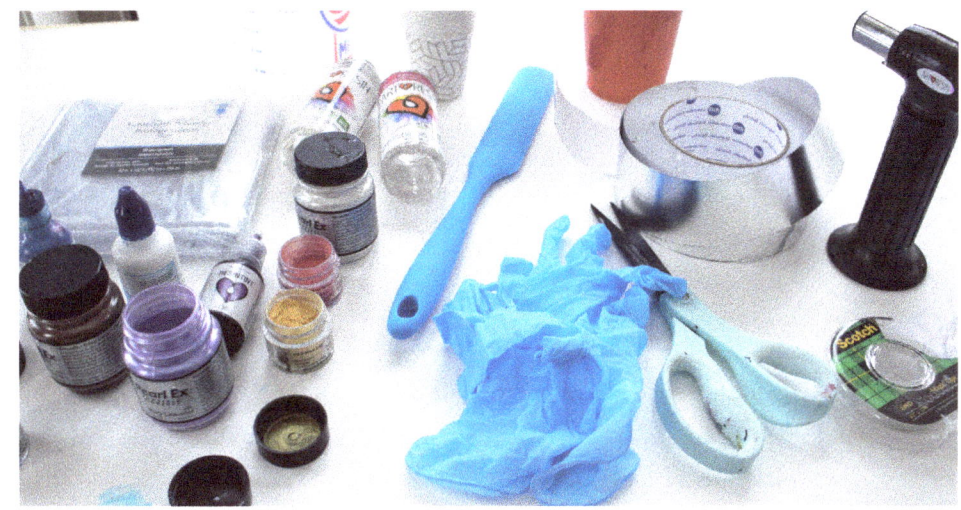

container or box to cover your artwork

oil-based pens and/or metallic markers

glitter and/or glass shards (optional)

small brush

matte gel medium

level (optional)

felt backing (optional)

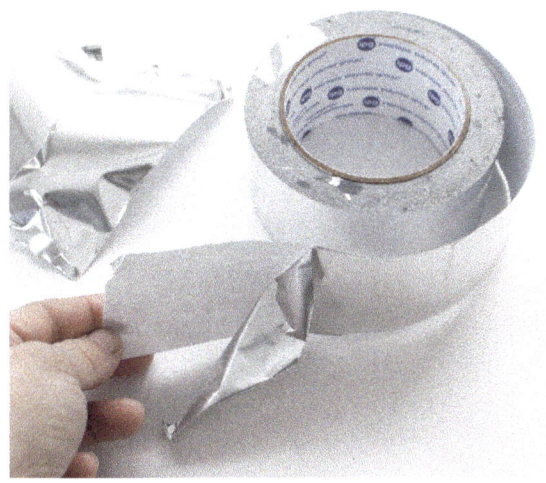

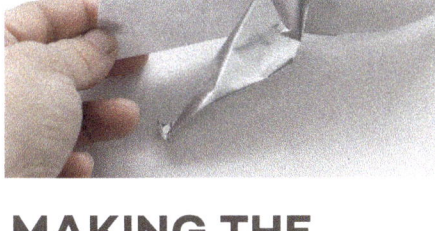

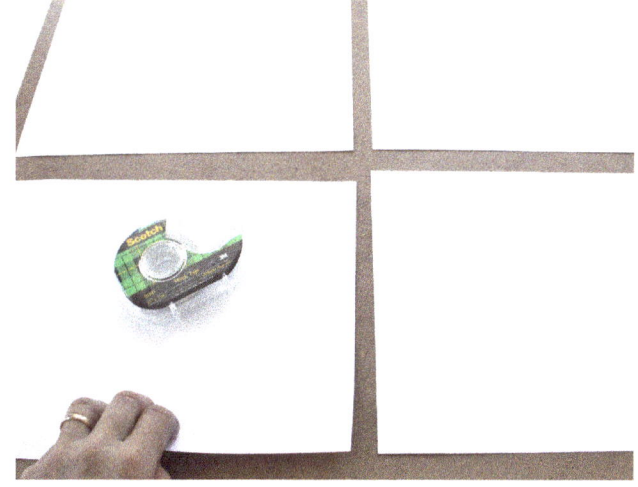

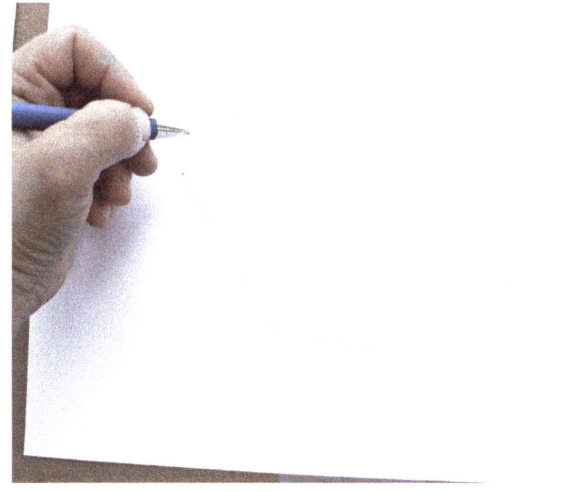

MAKING THE AGATE SLICE

1. The key to creating these amazing realistic shapes is all in the tape. Aluminum siding tape is a very sticky, sturdy, thick foil with a paper backing that you peel off. This type of tape can be found in most building supply stores. It is perfect for manipulating and creating shapes that hold their form. Regular tapes will not work for this, so don't try using duct tape or painter's tape because you will be disappointed with the outcome.

2. Start with a solid, flat sturdy work surface on which to create your molds. Cover the surface with a few pieces of white paper and tape in place. Begin drawing some shapes you'd like to create.

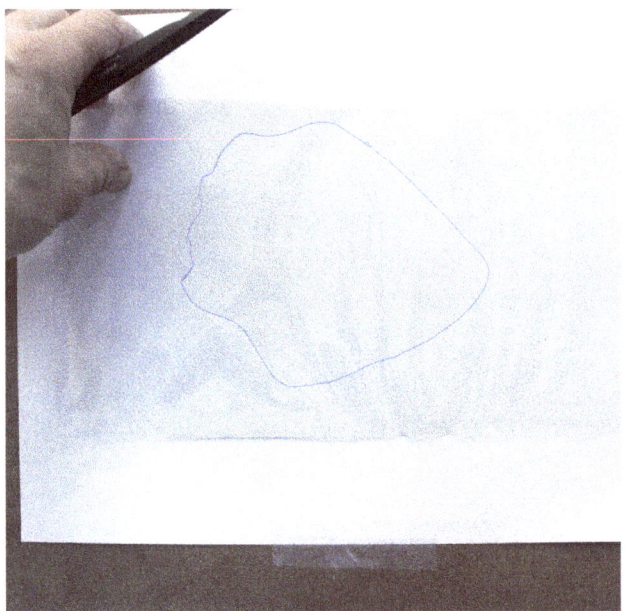 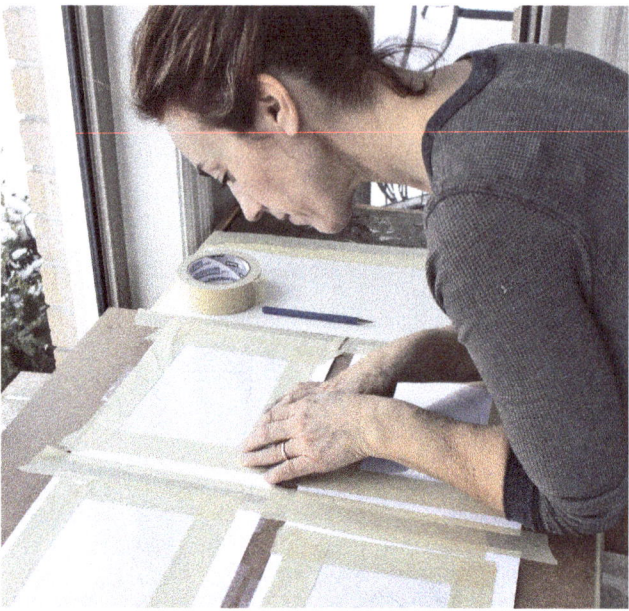

3. Cut your thick plastic shower curtain or tablecloth into large enough pieces to cover the drawing and paper itself. Tape the plastic in place, making sure it is pulled taut and is smooth without any bumps or wrinkles.

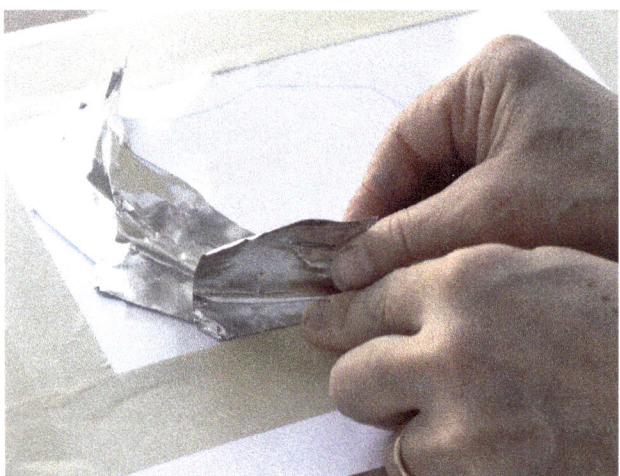 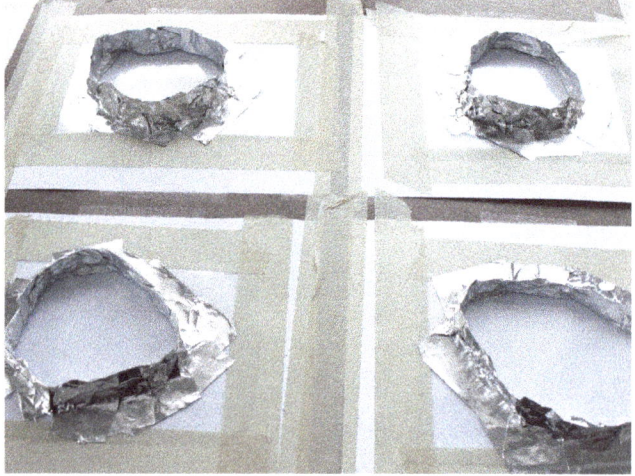

4. Begin cutting the aluminum tape into manageable pieces, remove the backing, and stick onto your plastic, following the lines of your drawn shape. Make sure there is a flap or folded piece of the tape sticking down. Continue to overlap piece by piece until your shape has been created. Make sure the tape does not have any gaps or areas where resin can escape. You don't want that! (Have a glue gun handy to cover up leaks if this should occur.)

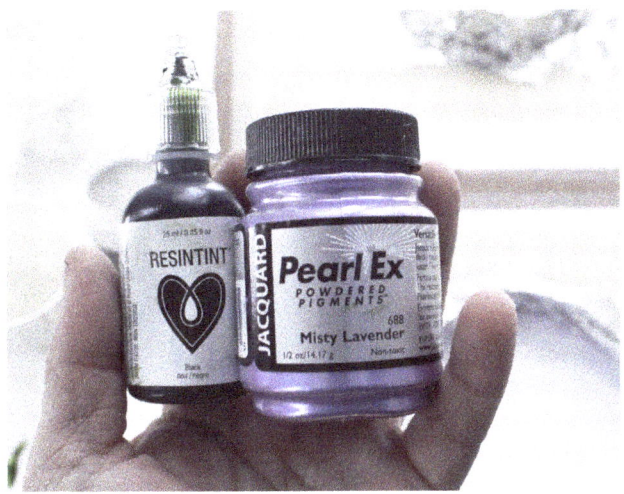

5. Resin can be colored in several different ways, including resin tints (liquid form) and mica powders (ground pigments). The amount of tint or powder you add will determine how translucent or opaque the resin becomes. Metallic mica powders produce a shiny look in resin. Do be careful not to add too much color, however. As a rule of thumb do NOT add more than 10 percent of colorant to resin because it may not cure properly.

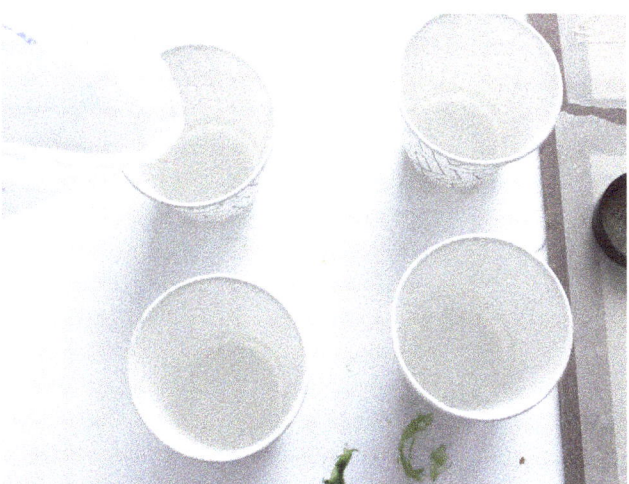

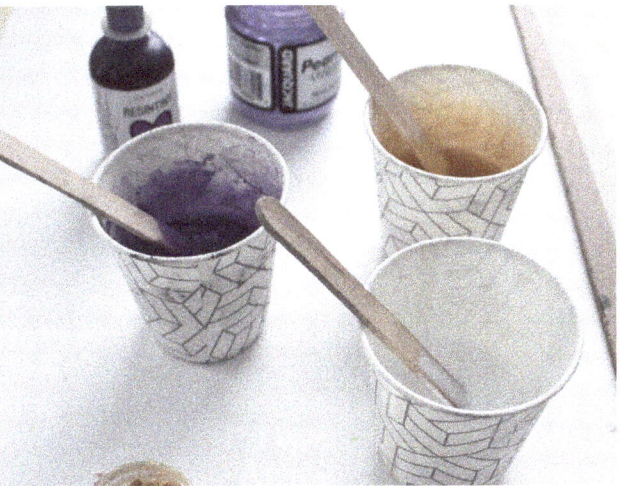

6. Mix your resin and pour into several other smaller containers to which you will add mica powder or resin tint. Use craft sticks to stir the colored resin. Make sure you mix mica powders well to avoid miniature powder bumps in your cured resin.

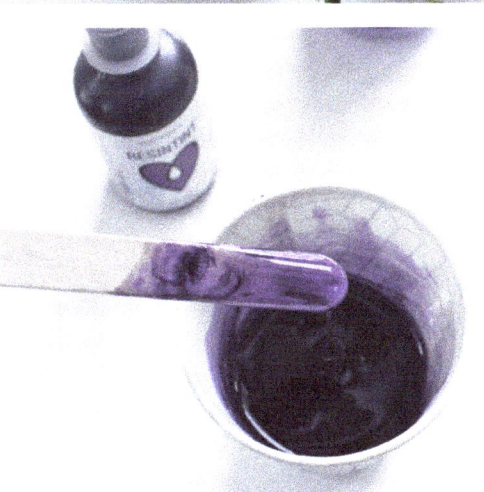

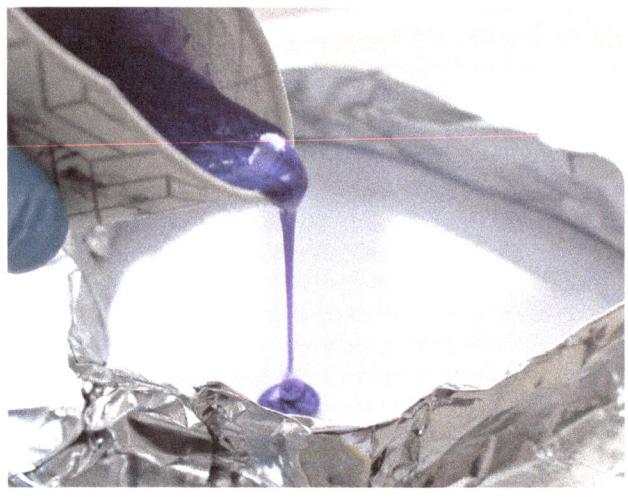 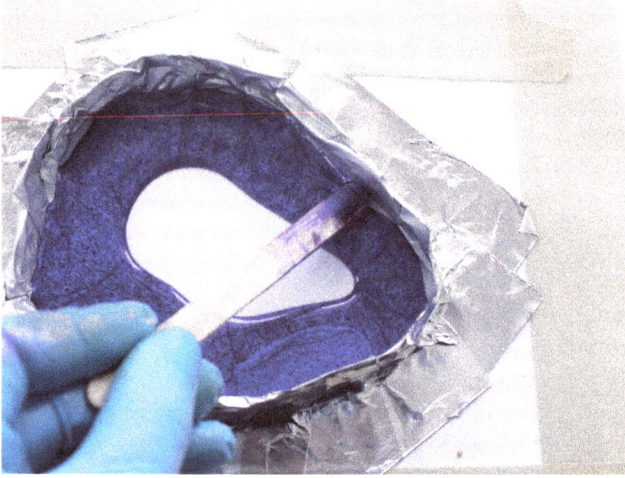

7. With your first color, pour around the inside edge with a bead of resin. Use a craft stick to move and manipulate the resin if needed.

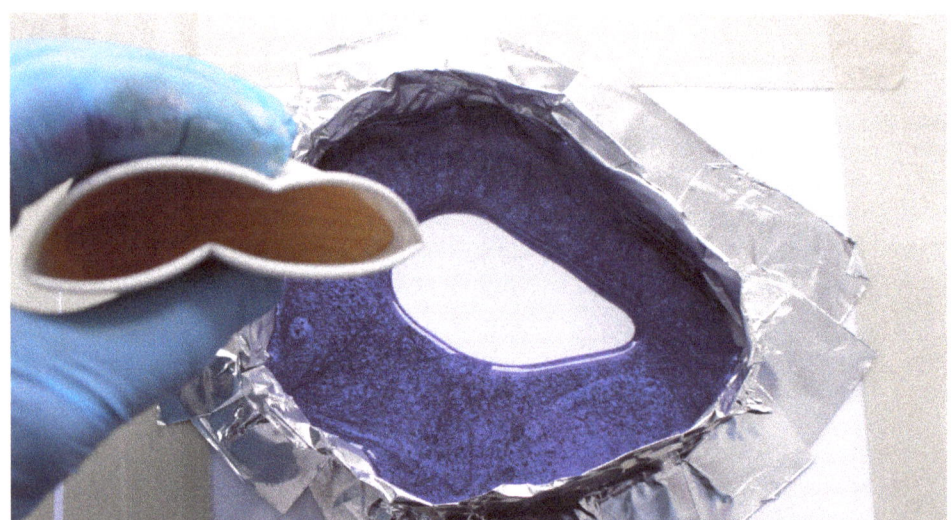

8. Continue the same process, following around the inside of the bead you just poured with a second color. You can repeat again with another color. Remember to pay attention the level of the resin in the mold, taking care not to make it too thick.

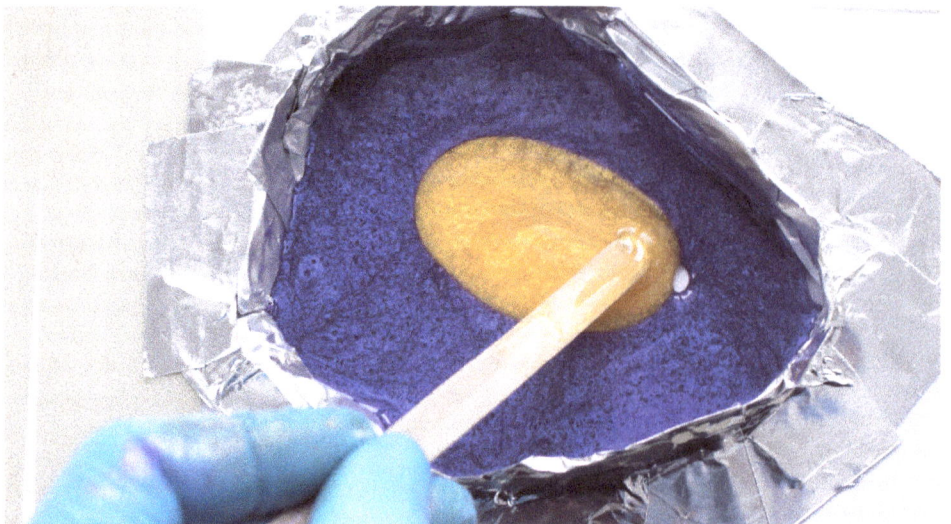

TIP

Do not fill more than halfway up the well or the resin will be too thick and may not cure properly.

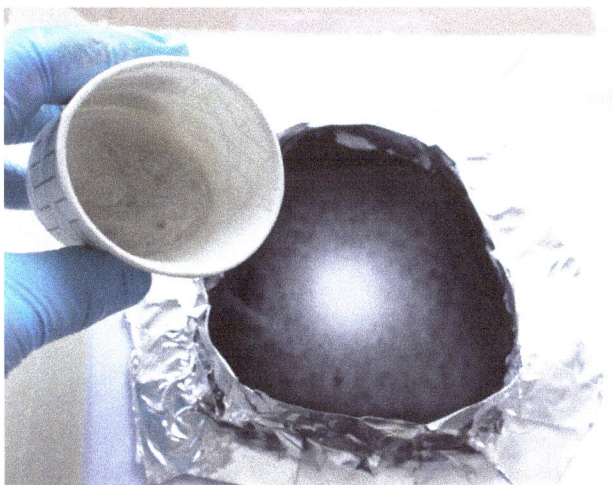 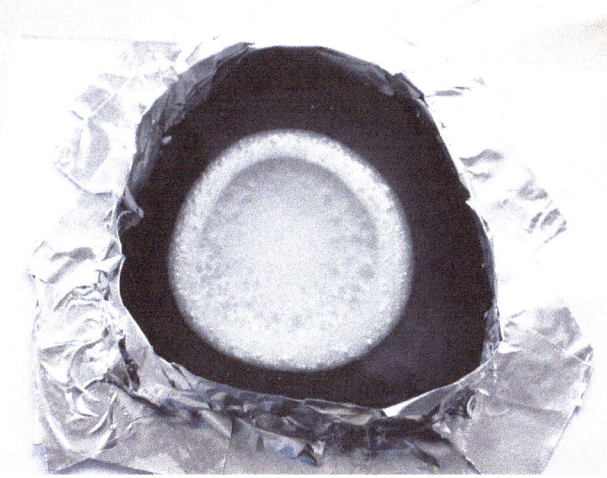

9. For your next shape, try pouring in the center first, which pushes the colored resin layers out. Make the last center pour clear resin so you can see through the center of the coaster when it's cured.

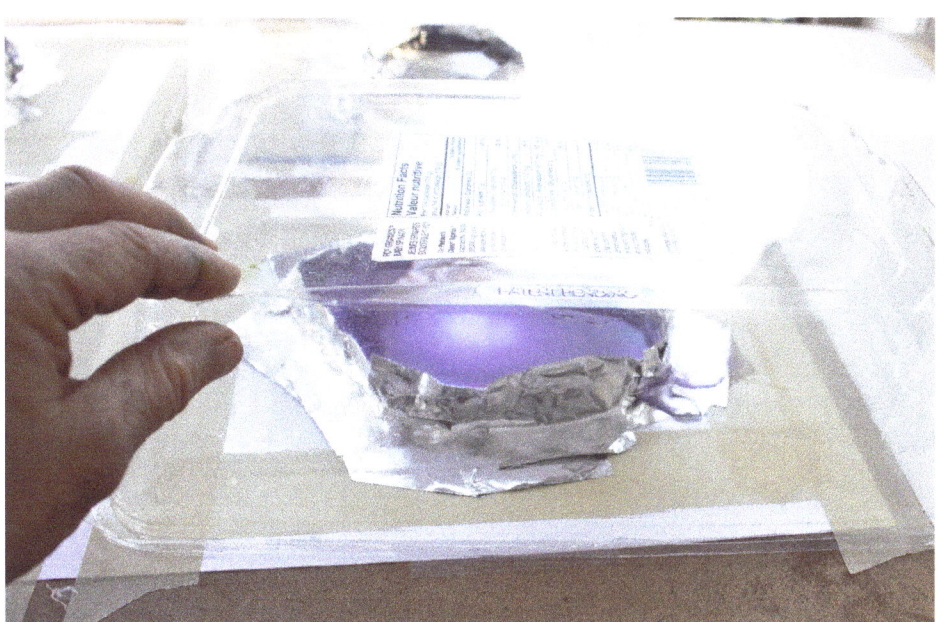

TIP

For more information about the torching process, review the MOD Mini resin stage process or check out the Resin FAQ/ Troubleshooting section in this book for tips and tricks.

10. Resin continues to transform as it cures, so what you started out with may not always be how it looks in the end. You do, however, have some control over the resin depending on the stage at which you manipulate it and the brand you use. With Art Resin, you have approximately 45 minutes of working time. If you try swirling a design around with a toothpick as soon as the resin is poured, you will not maintain that effect. But if you leave the resin a bit longer and create some swirly lines closer to the end of the working time, you will have a greater chance of ending up with more detail in the cured piece. Resin is all trial and error, so have fun experimenting and see what transforms!

11. Before allowing your shapes to cure, torch lightly to remove bubbles (if needed) and check for leaks around the base. If leaks occur, use a glue gun to seal up the area. Cover with a container to prevent dust from settling and leave for at least 12 hours.

REMOVING FOIL TAPE

Peel enough of the tape so that you can pry the shape away and off the plastic backing. Begin removing foil from the edges. The resin shape itself will still be a bit pliable, which is okay beacuse it makes it easier to remove the foil. You will need a bit of patience for this step, as the tape will be stuck into some of the crevices. Use a toothpick or pin if necessary to help remove it from the pesky areas. Once all the foil is removed you will be left with an agate slice form with jagged edges. Place the shapes flat on your work surface and allow them to cure until you can no longer bend the resin.

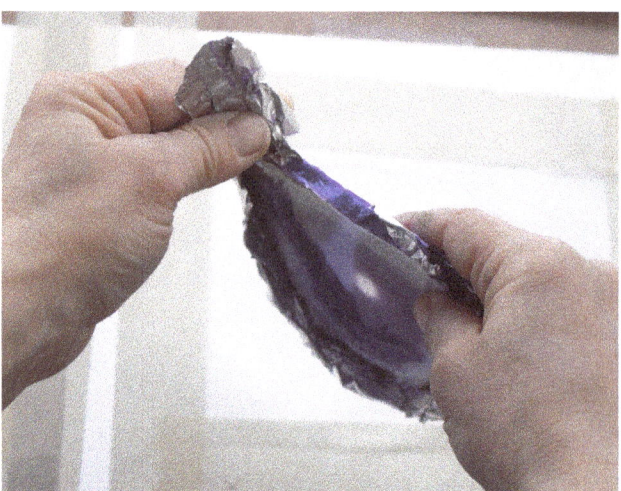

NOTE

It is very important that you remove the foil tape BEFORE the resin fully cures or you will have difficulty removing it. Once the resin is at a fully hardened state (24 hours for Art Resin), you can't remove it!

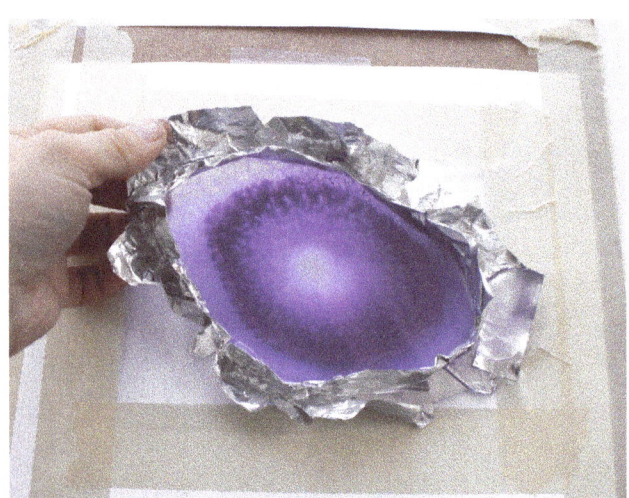

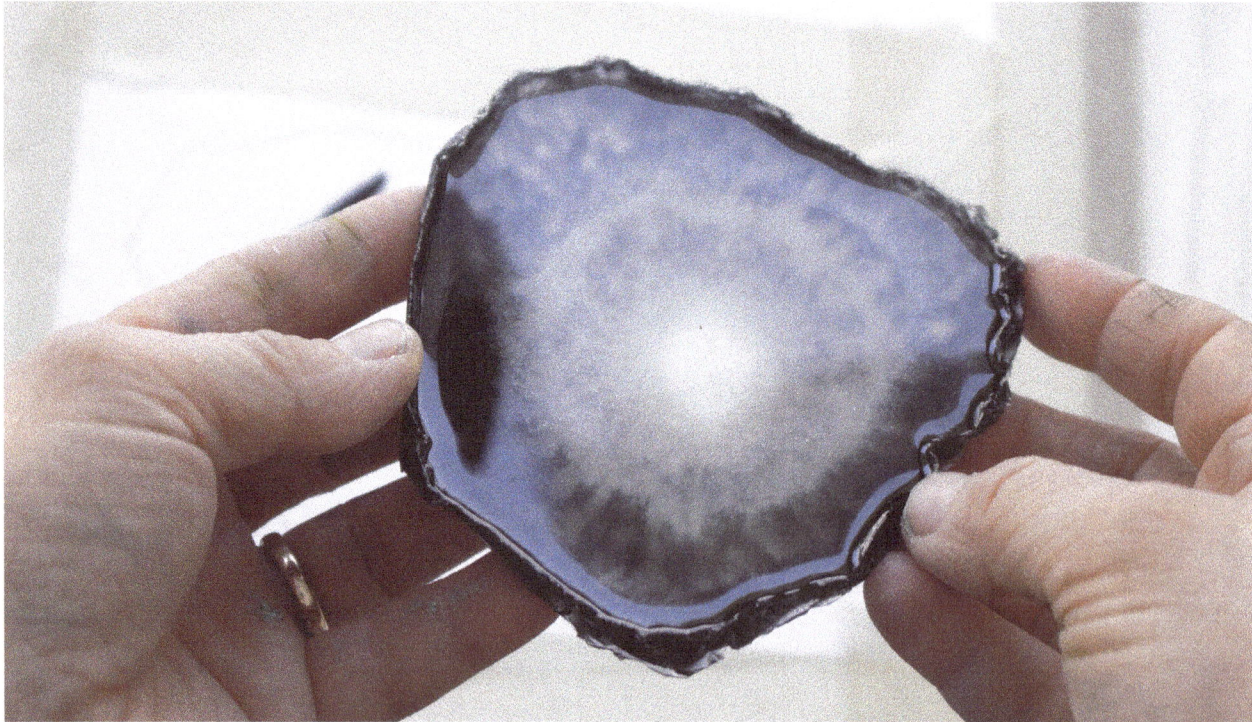

EMBELLISHING YOUR AGATE SHAPES

Next, have fun embellishing your agate slices with colored, oil-based pens and metallic markers. Finish your edges with either gold gilding paint, metallic acrylic paint, or even gold marker. At this point you may wish to leave your agate shape as is with the rough natural edge. For a finish with a smoother edge, coat with another layer of resin.

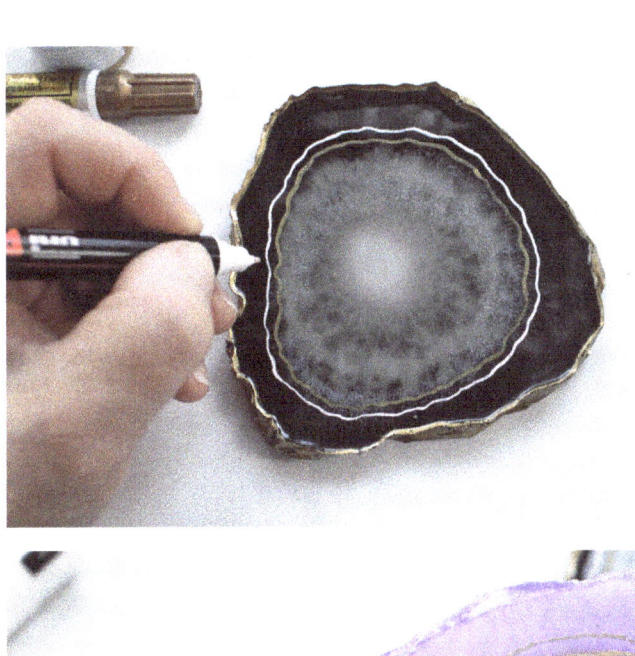

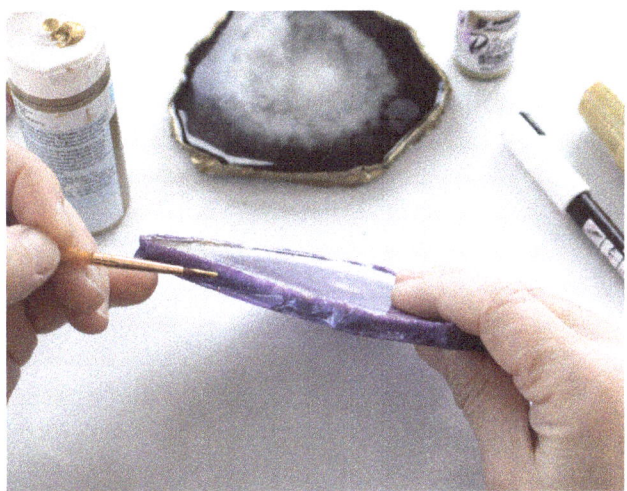

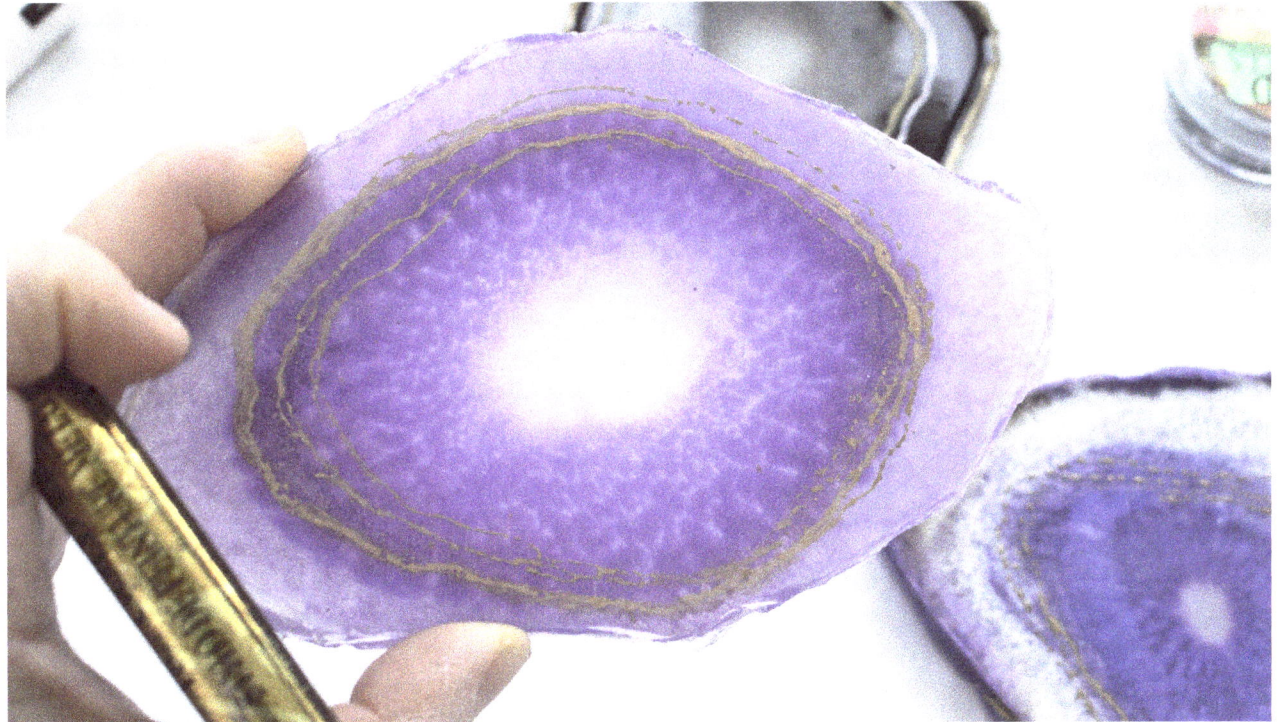

OPTIONAL SECOND COAT OF RESIN

If you want a smooth edge with no ridges, you will need to add a second layer of resin. At this point you may also choose to embellish further with additional glass remnants or glitter before applying your second coat. Use a small brush with some matte gel medium or any clear drying adhesive to secure particles in place.

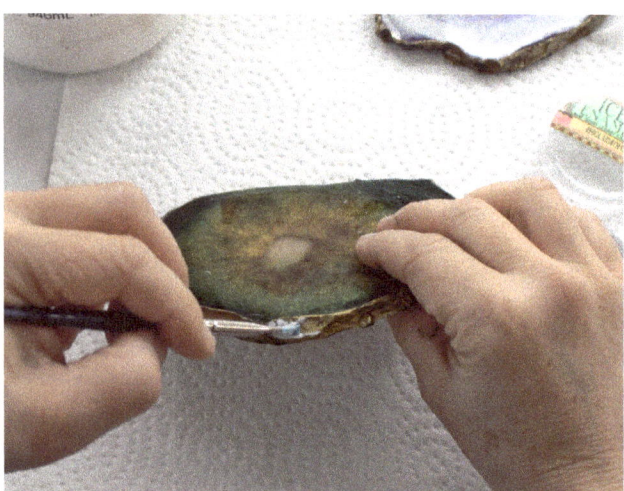
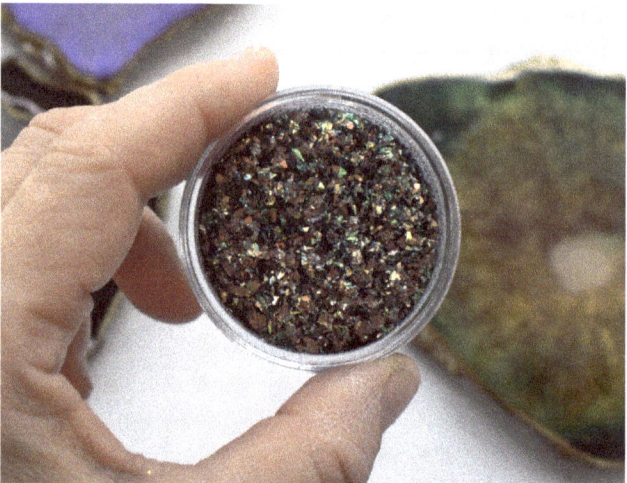

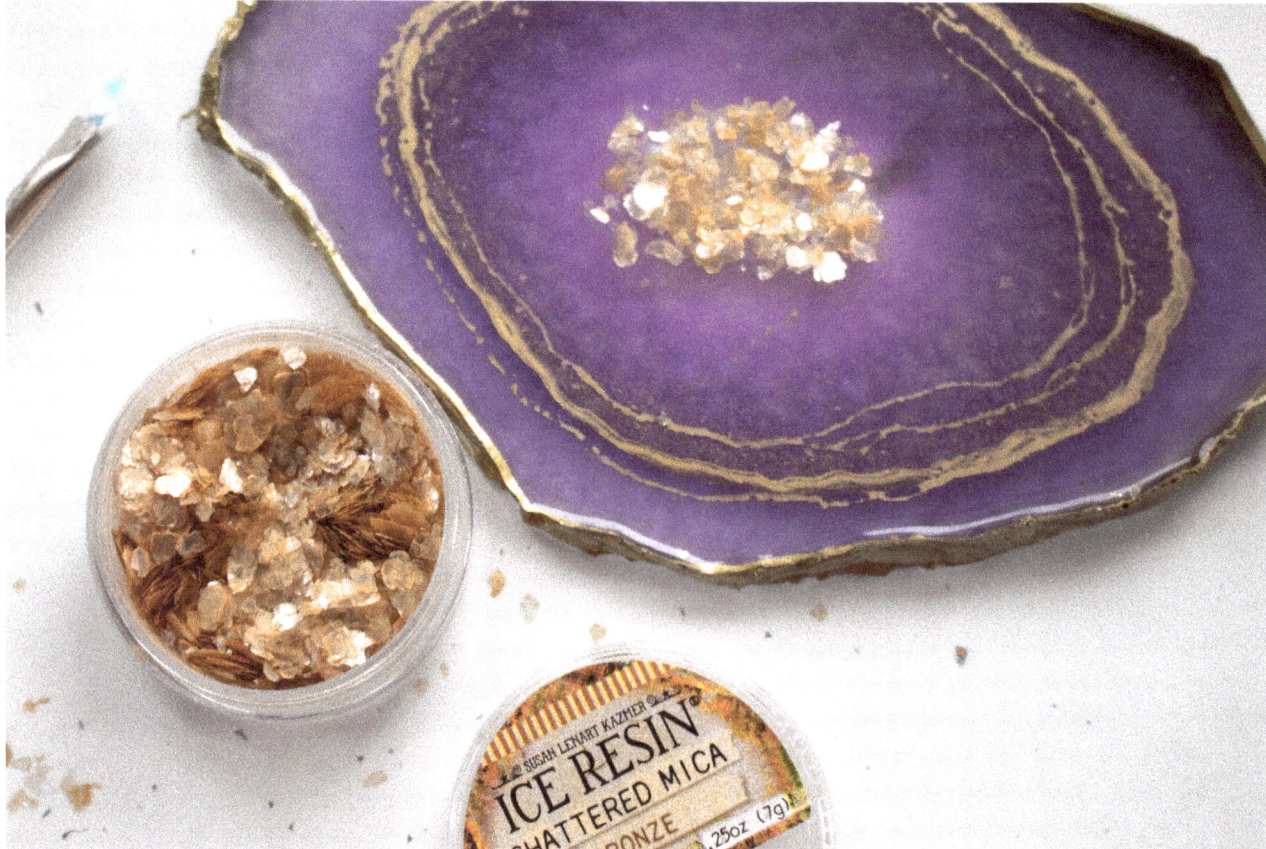

TIP

If you are planning to apply a second layer of resin, always handle your resin work with gloves. When you handle resin with bare hands, the natural oils from your skin can transfer to the resin, leaving a residue on the surface. This causes the second coat of resin to repel and not adhere properly. If you have handled the cured resin without gloves, lightly wipe the surface with a warm, wet paper towel and some mild dish soap. You may also lightly sand the surface.

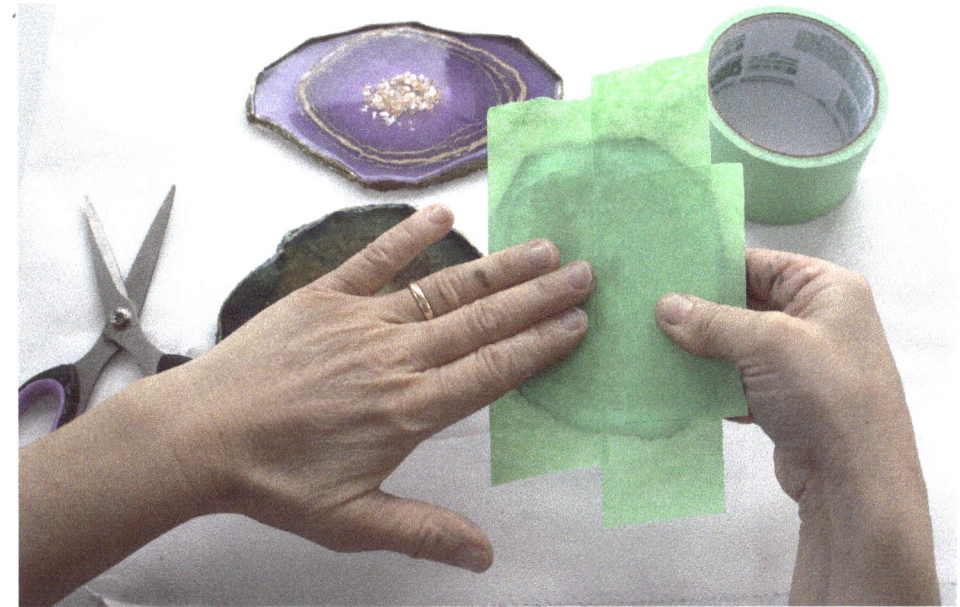

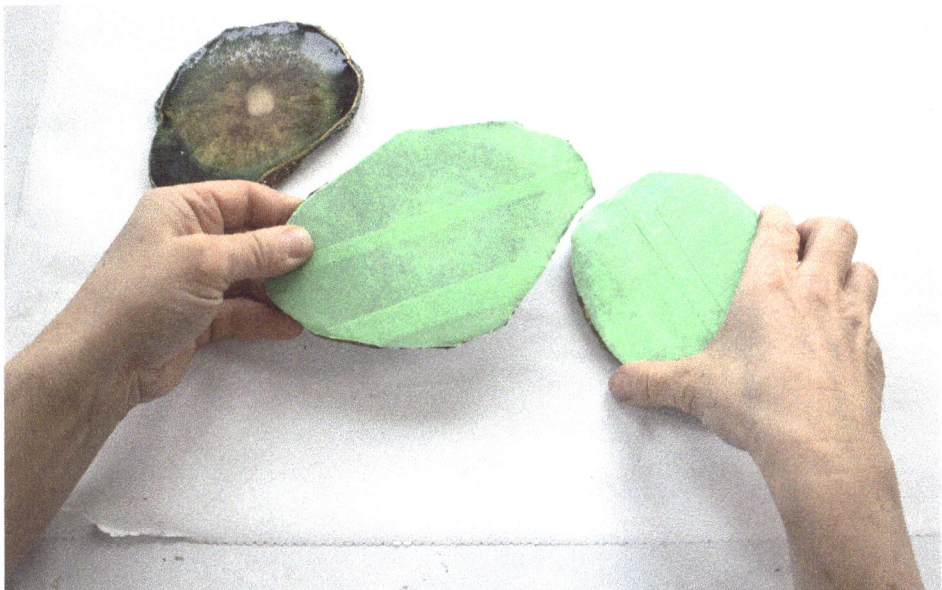

1. Before applying the second coat of resin, apply tape to the backside of the coaster and trim the edges flush. This will prevent resin drips from curing underneath.

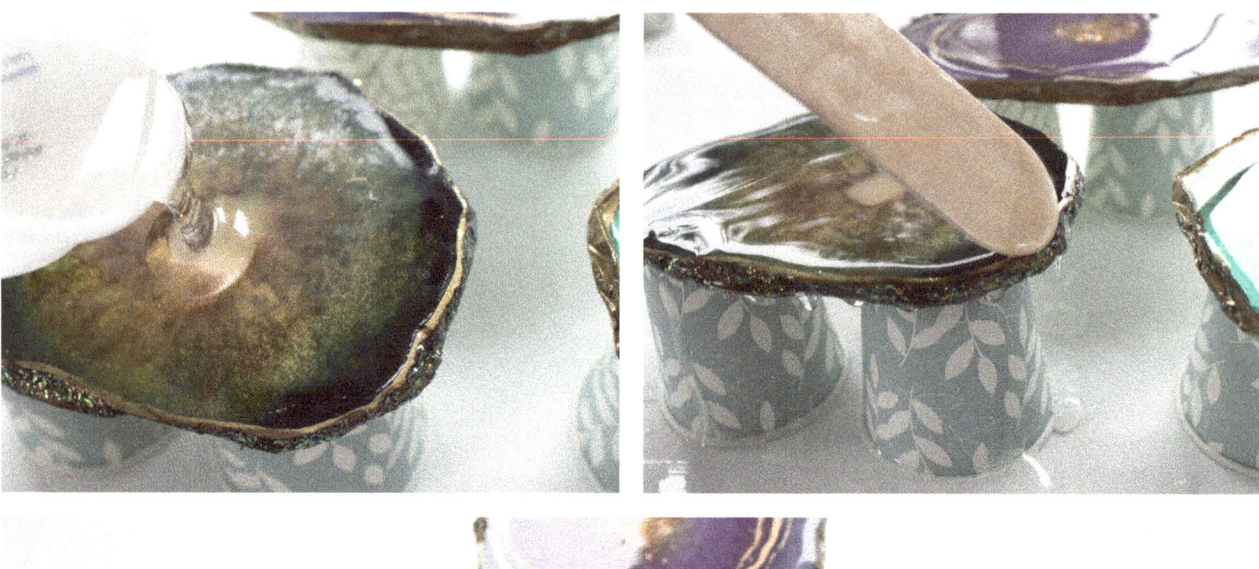

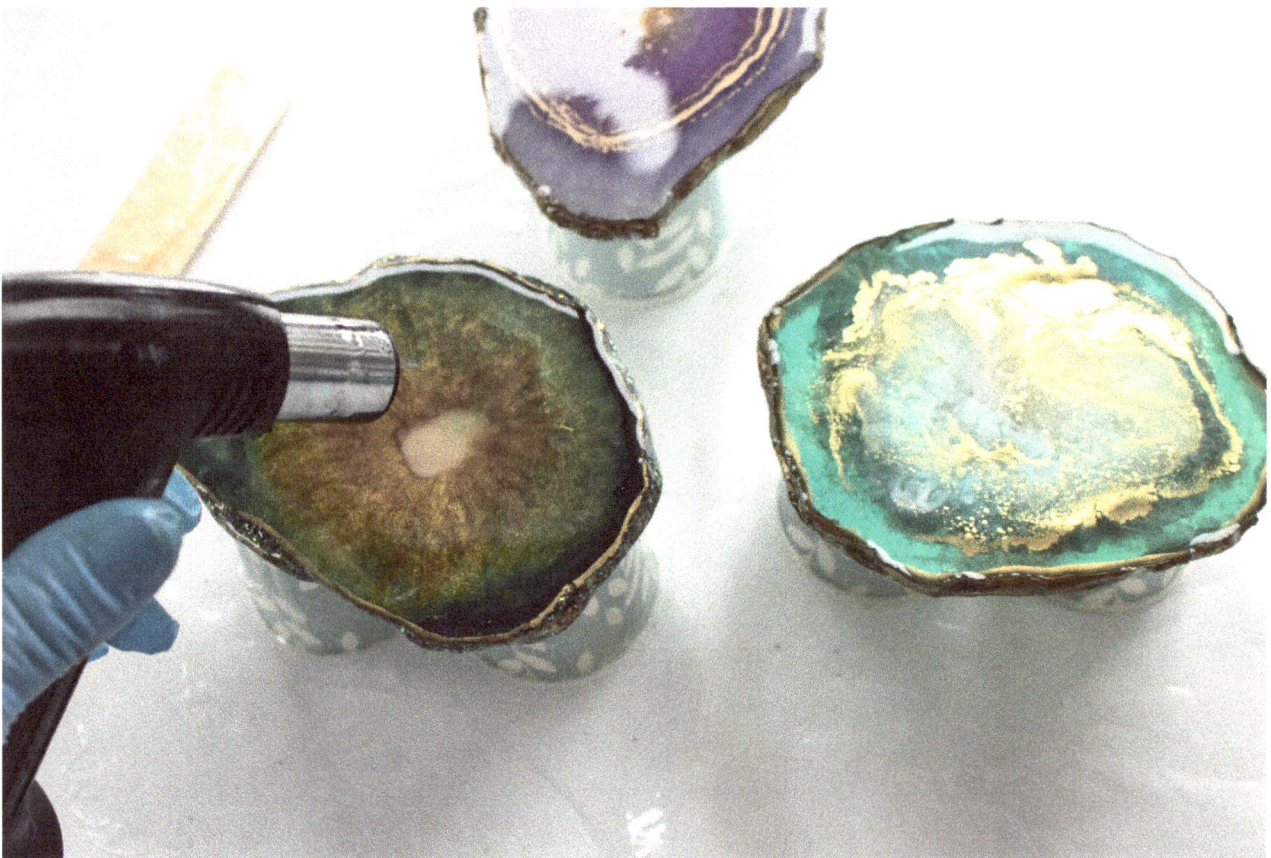

2. Set your agate slices on paper cups atop a protected surface, ensuring they are level before covering with resin. Smooth over evenly with a craft stick, torch if needed, cover, and allow to cure overnight.

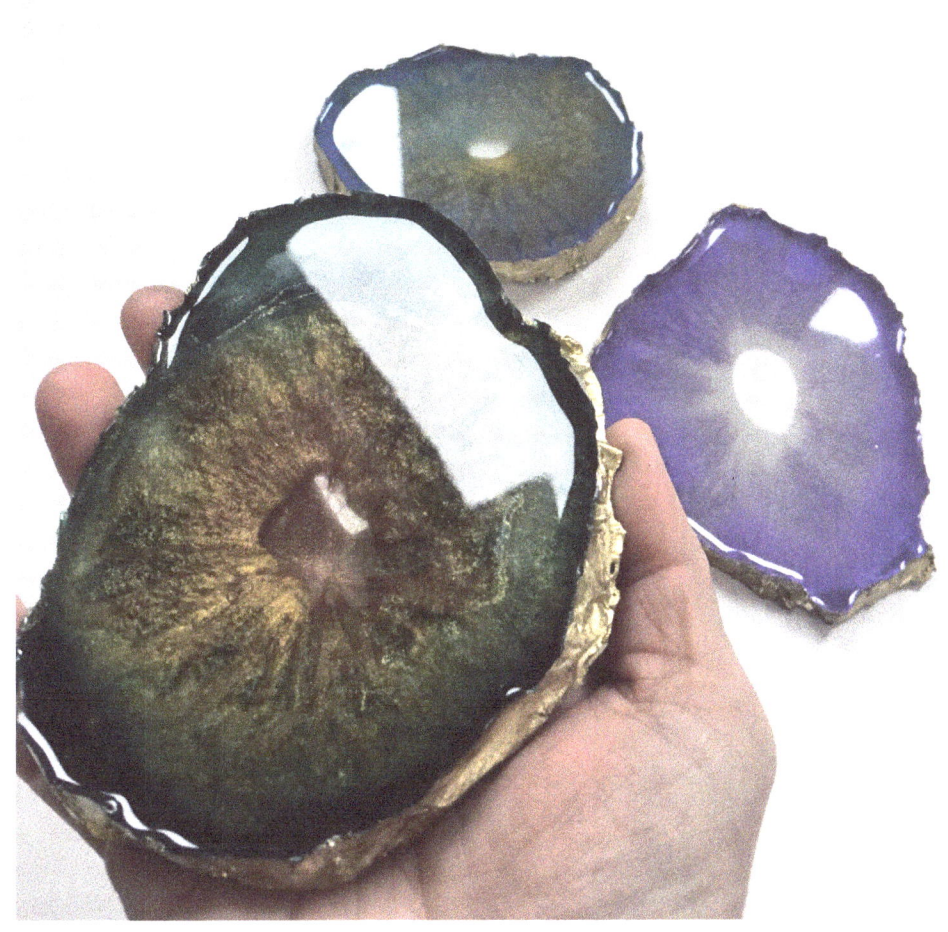

3. Remove the tape from the backside. For added protection, adhere either a cut-out felt backing or protective dot backers. Your coasters are ready for use!

TIP

Different resins do not hold up as well when hot beverages are placed directly on the surface. Check the brand of resin you're using to determine whether it's suitable for hot drinks. Art Resin can tolerate heat up to 120°F (49°C).

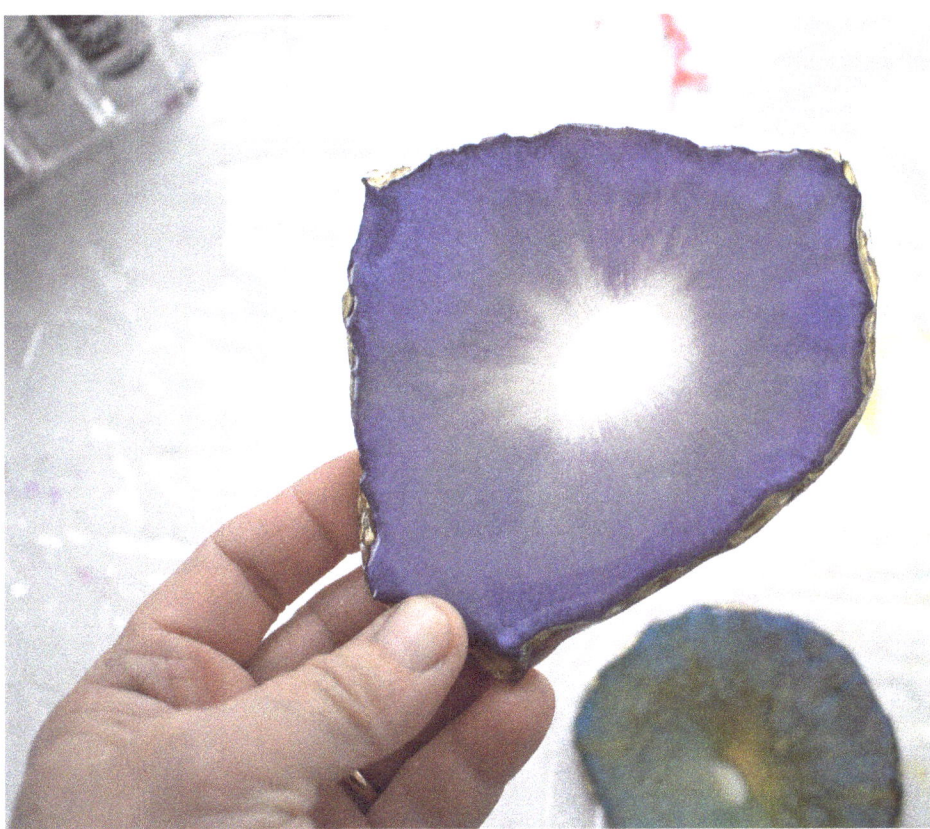

ALCOHOL INK CERAMIC COASTERS WITH TWO TECHNIQUES

Alcohol inks can be used on any nonporous surface, including ceramics. Inks can be dropped, painted, or blown around the surface to create interesting patterns, color, and detail. Tile is one of the easiest substrates on which to work with alcohol inks, primarily because you can easily wipe inks clean from the surface with isopropyl alcohol and start over again. Only a few colors are needed to produce vibrant results. In this project you will create two distinct styles using two different techniques. The results look as if you spent a lot of time, but they actually take mere minutes.

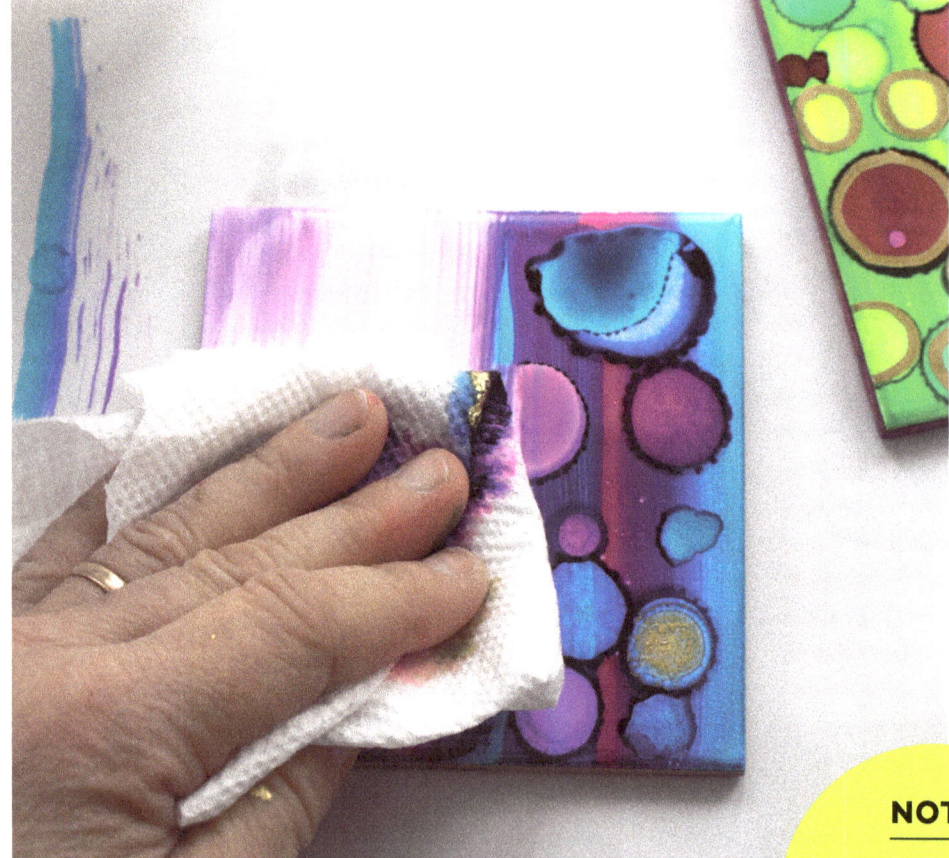

MATERIALS

tiles in any shape

paper towel

isopropyl alcohol or blending solution

alcohol inks

heat embossing tool

gold markers or gold embossing powder

sponge brush

spray sealant

UV protectant spray

NOTE

Follow safety recommendations as outlined in the beginning of the book.

TECHNIQUE 1
TILE COASTER

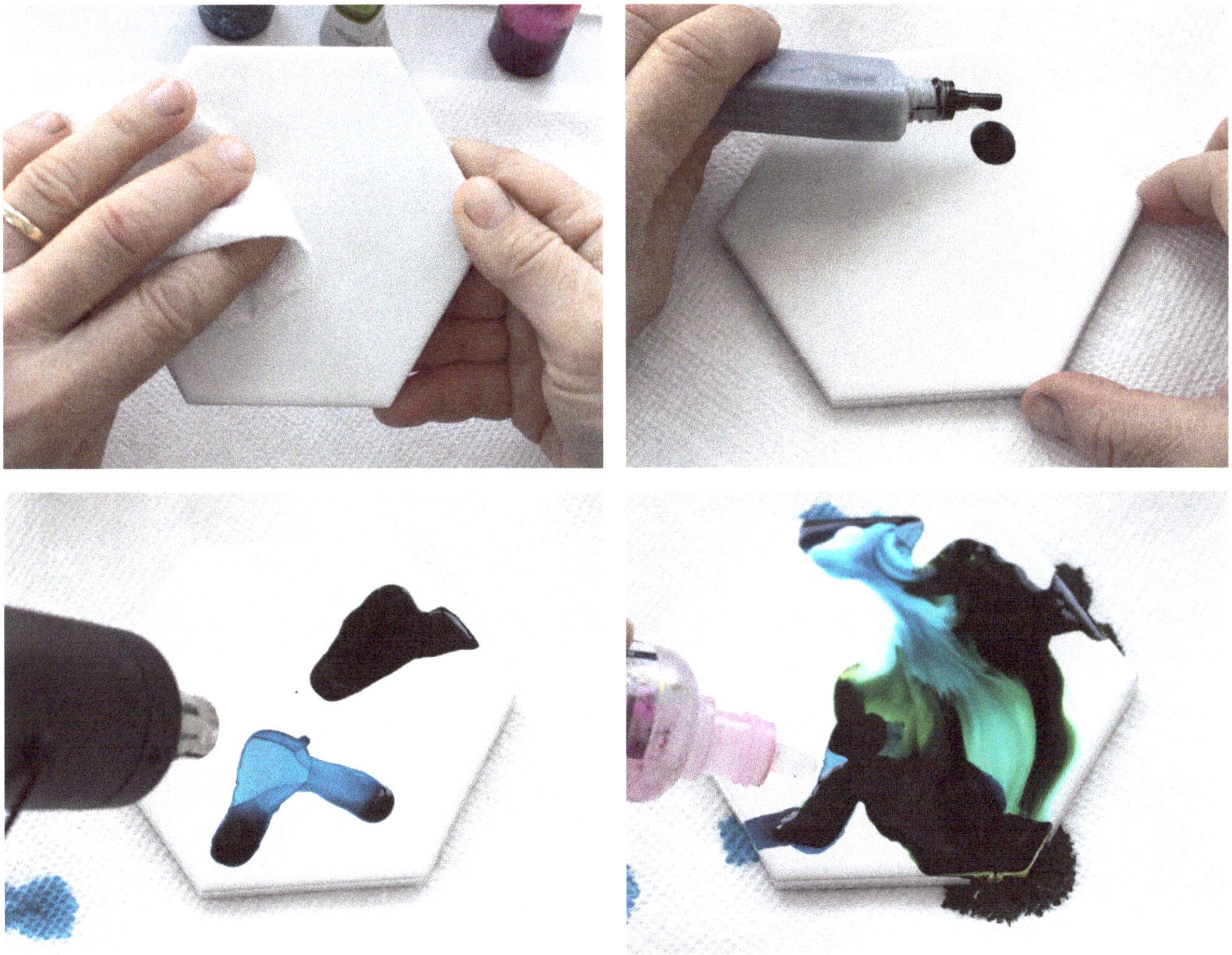

1. Begin by cleaning your tile with a paper towel and alcohol. Select two or three alcohol ink colors and add a few drops onto the tile, using the embossing tool to dryset the inks in place. Next, drop a generous amount of isopropyl alcohol or blending solution over the inks.

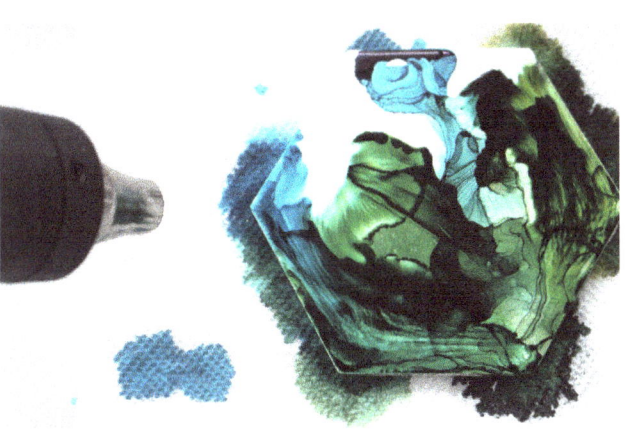

2. The key to this wispy ring-like technique is using an embossing heat tool instead of a hair dryer, which will blow the inks around without much control, even on a low setting. An embossing tool is easier to control with gentler heat and air than an actual heat gun, allowing you to create a wispy, ring-like effect.

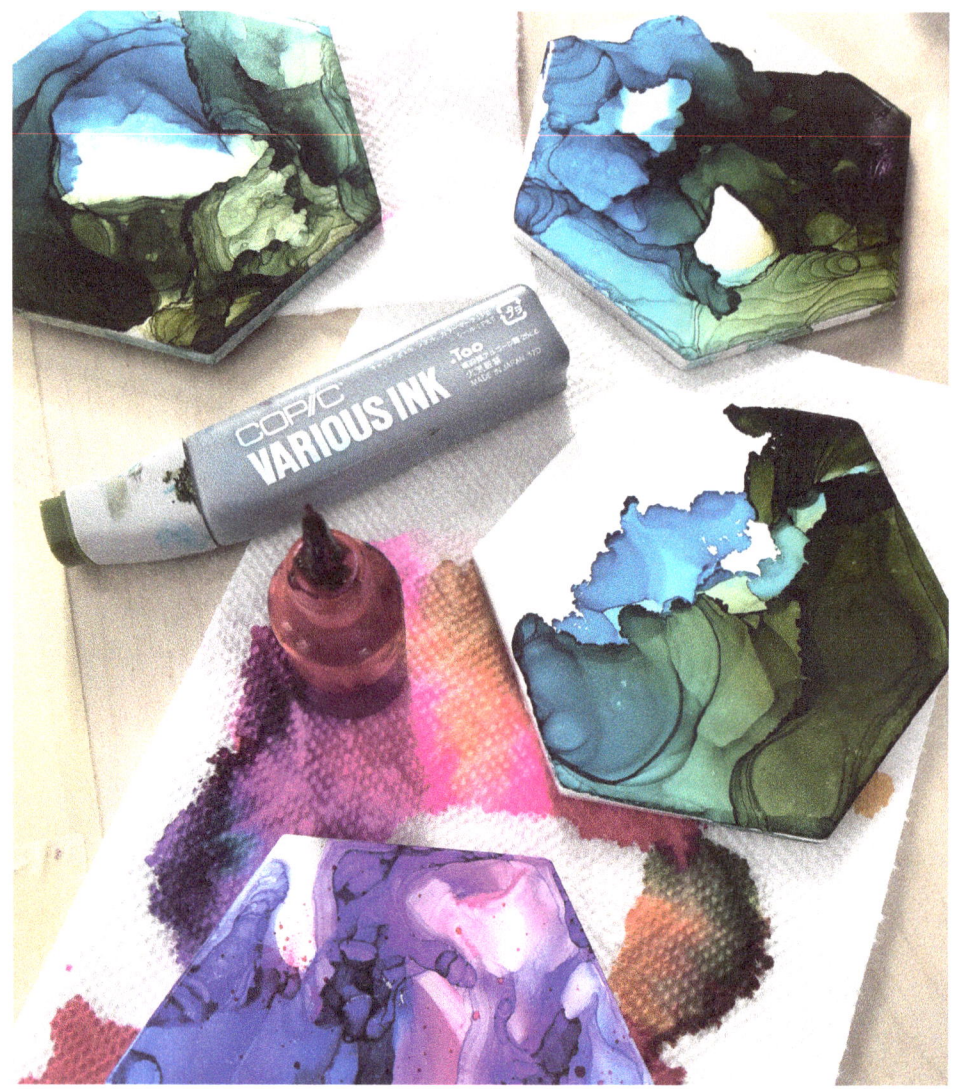

3. To create this effect, hold the tool slightly above the wet solution and, using a back-and-forth motion, dry the inks every few seconds. The inks will act like a pool of water, drying into ring-like patterns. With this technique, the inks will dry in 10 to 20 seconds! Add more solution if desired and repeat the process. Try different colors on several different tiles. That's it!

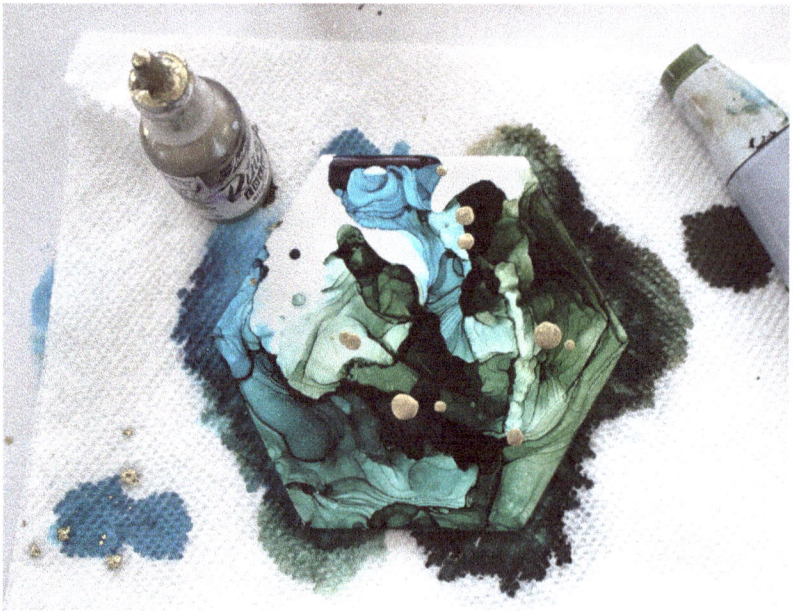

4. Finish your tiles with either gold metallic alcohol ink drops, or heat emboss with gold powder.

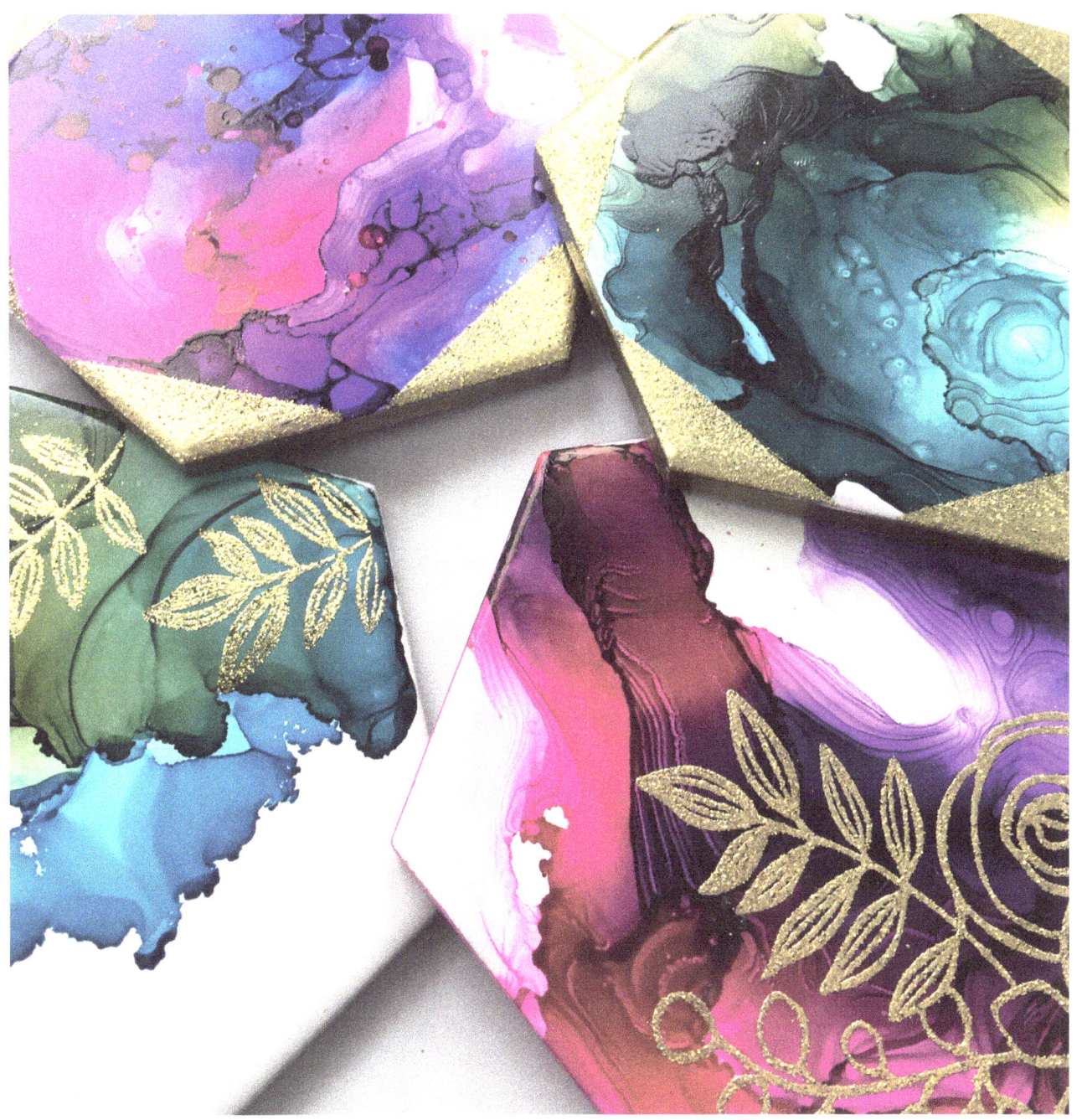

Take your tiles to the next level by adding heat-embossed stamping. Visit my YouTube channel for the full embossing on alcohol ink video tutorial. Visit www.youtube.com/janemonteith.

TILE COASTER

This technique also takes a few minutes to create. Tiles can be embellished with gold metallic oil-based markers for added texture and appeal.

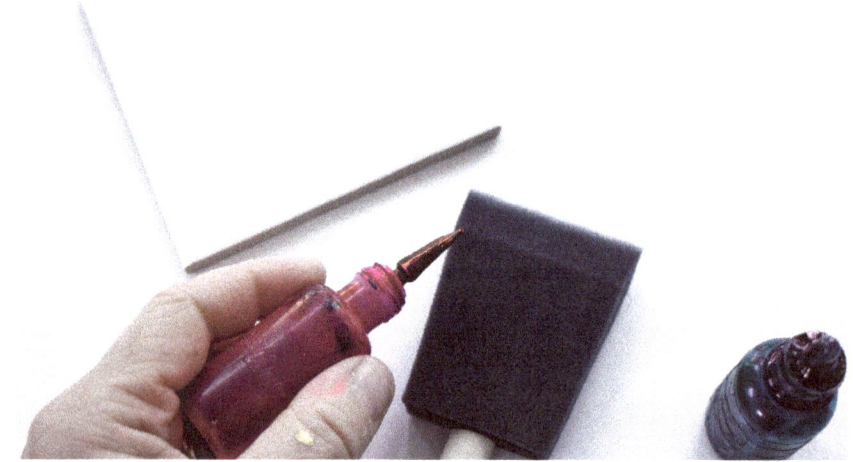

1. Begin by cleaning off your tile with paper towel and isopropyl alcohol. With a sponge brush, squeeze several colors of ink across the top.

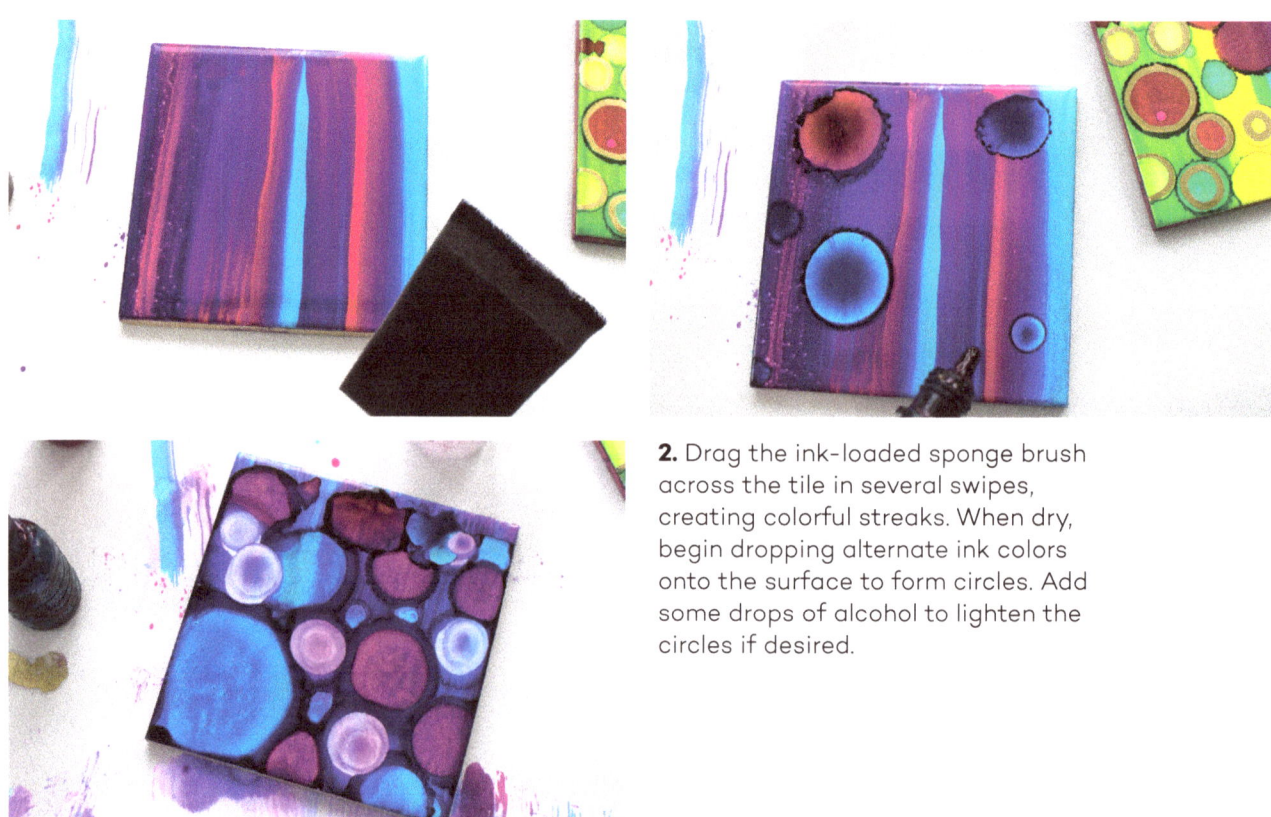

2. Drag the ink-loaded sponge brush across the tile in several swipes, creating colorful streaks. When dry, begin dropping alternate ink colors onto the surface to form circles. Add some drops of alcohol to lighten the circles if desired.

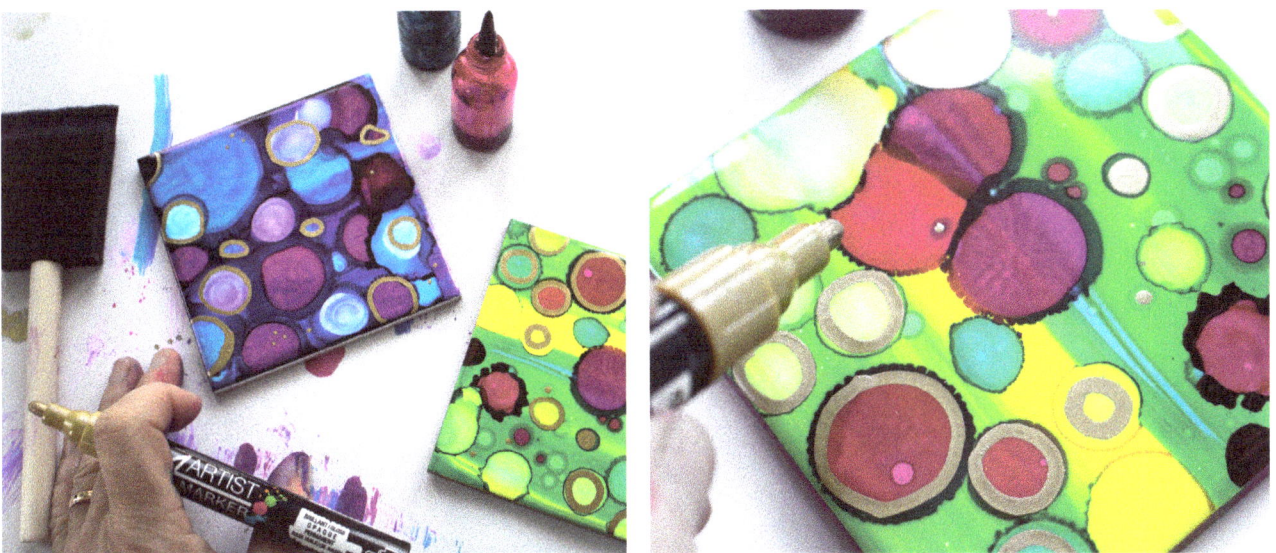

3. Allow the tiles to dry before adding gold markings.

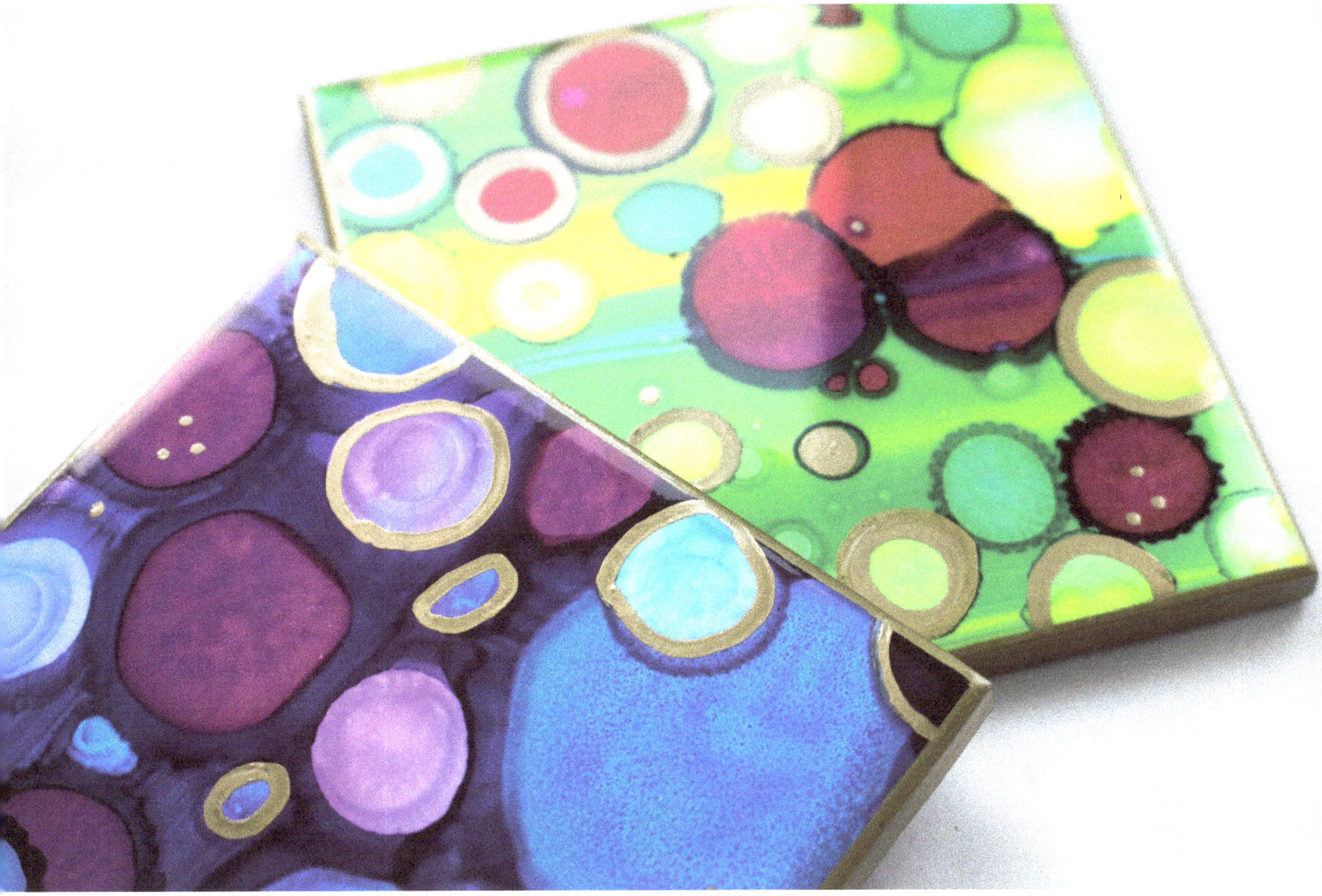

4. To protect your tiles, spray with a sealant to prevent bleed and a UV spray to prevent fading, as mentioned in project 3. You can also add resin to your tile coasters and finish the back with cut-out felt (see project 5).

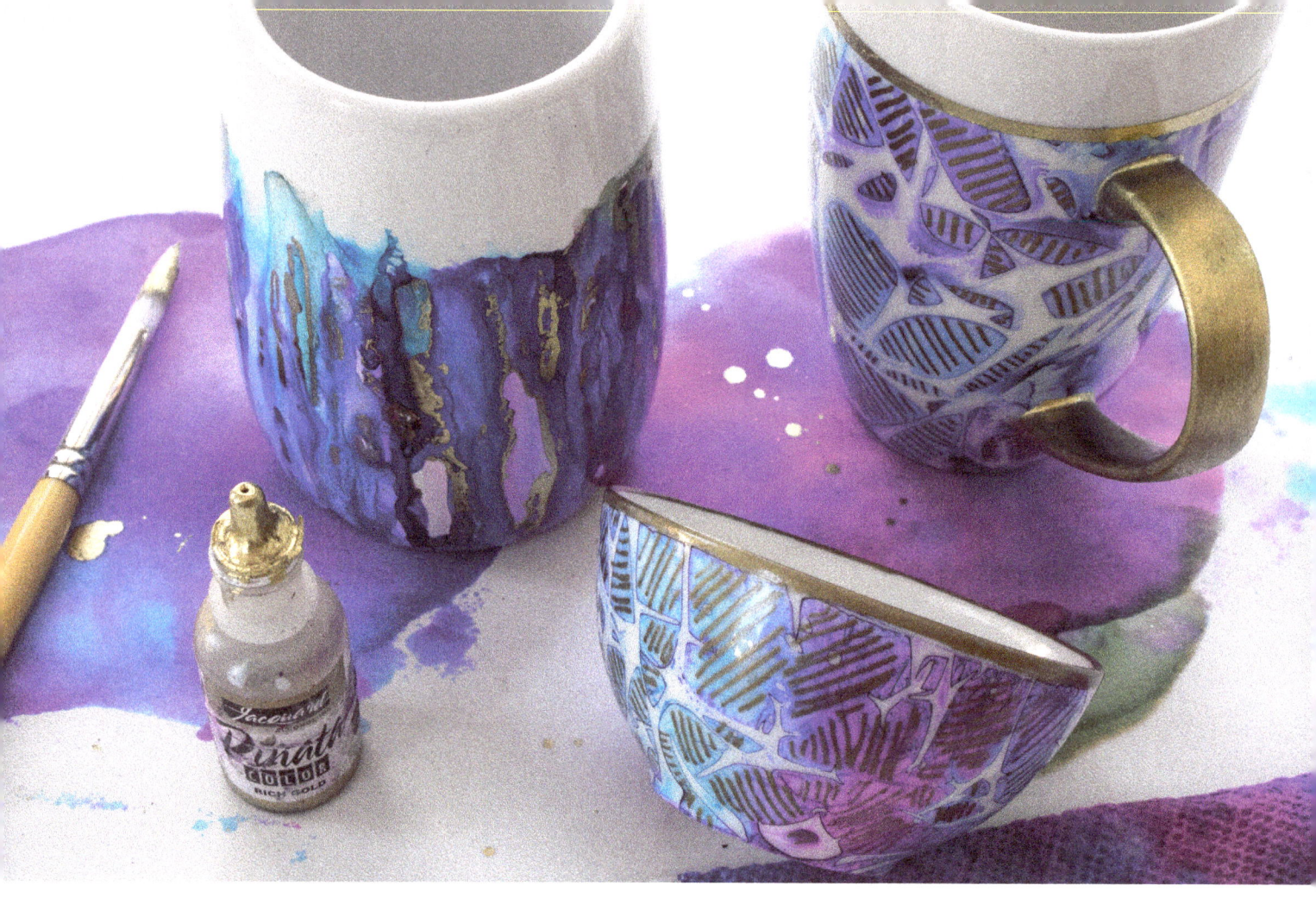

PROJECT 7

FLUID CERAMIC VESSEL AND MUG WITH TWO TECHNIQUES

I love these fun and simple techniques that can be used to create decorative pieces using only three colors. Each technique is slightly different and the results will leave your friends wondering how you made them. They make great gifts too!

MATERIALS

black garbage bag or plastic tablecloth to protect surface area

protective gloves

white ceramic vessel and mug

paper towel

alcohol inks, including gold metallic (shown here: Jacquard Piñata Senorita Magenta, Baja Blue, and Rich Gold)

blending solution or isopropyl alcohol

sponge brush

gold metallic oil-based marker (used here: Pebeo 4 Artist Marker)

plastic cling wrap

masking tape

small paintbrush

Krylon Kamar sealant

Krylon or Golden UV Archival spray (optional)

TECHNIQUE 1
CERAMIC VESSEL

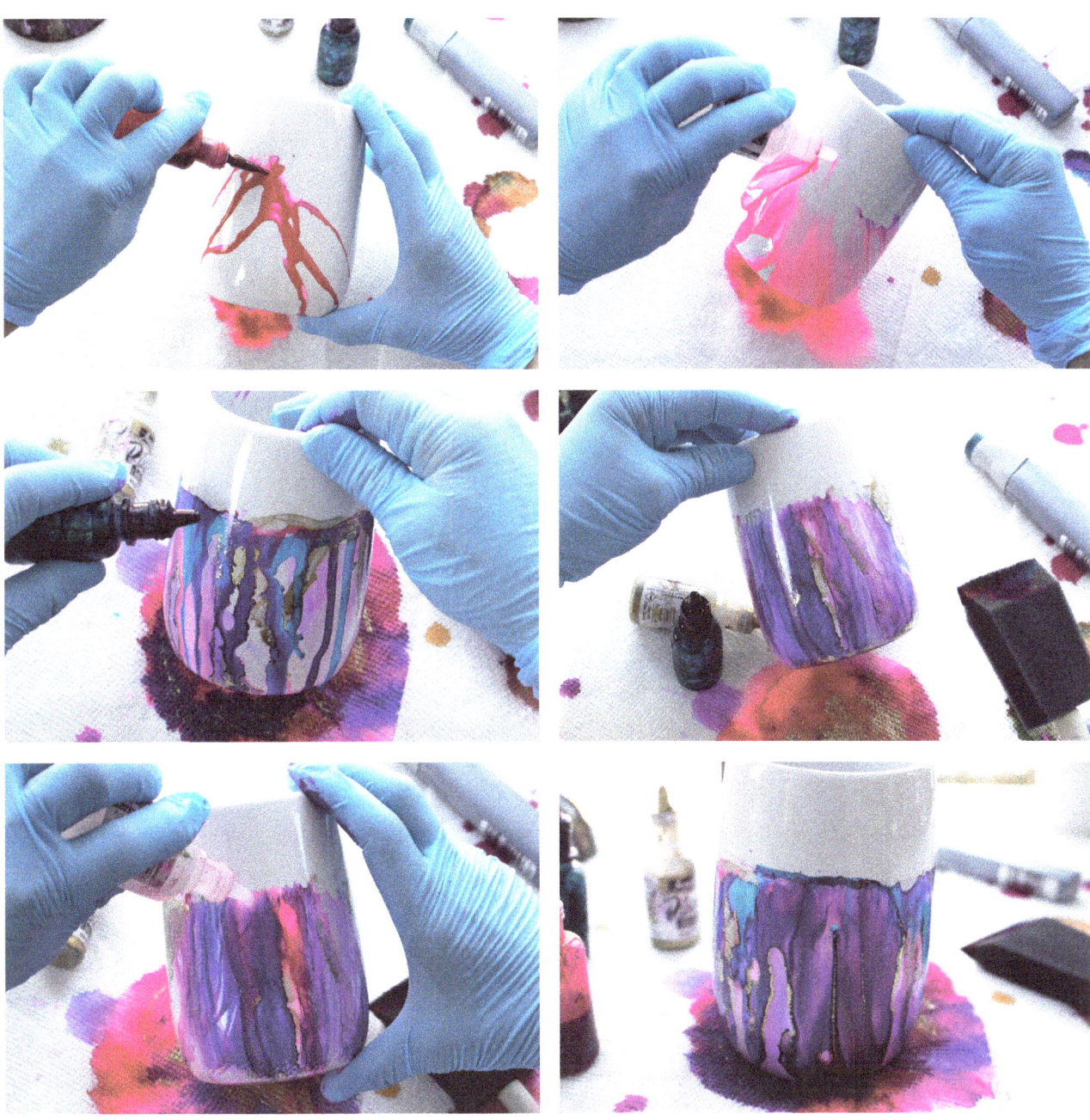

1. Cover your work surface with a black plastic garbage bag or plastic tablecloth. Put on gloves and hold the ceramic container over a paper towel. Turn the vessel at an angle and begin dripping one color from the top third of the ceramic container down. Continue around the entire vessel. Allow the excess drips to fall onto the paper towel. Drop some alcohol or blending solution in the same fashion around the container to soften and blend the ink. Continue alternating with the second color and then gold. Brush down once with a sponge brush to again blend a few of the colors together. Keep repeating until you're happy with the colors. Wipe off any smudge marks with a paper towel and alcohol.

TECHNIQUE 2
TEXTURED CERAMIC MUG

This is one of my all-time favorite alcohol ink techniques using cling wrap. It's extremely easy and the results always make me smile. You can add extra touches of gold to really bring a wow factor to this finished piece. This technique looks great on trinket bowls too.

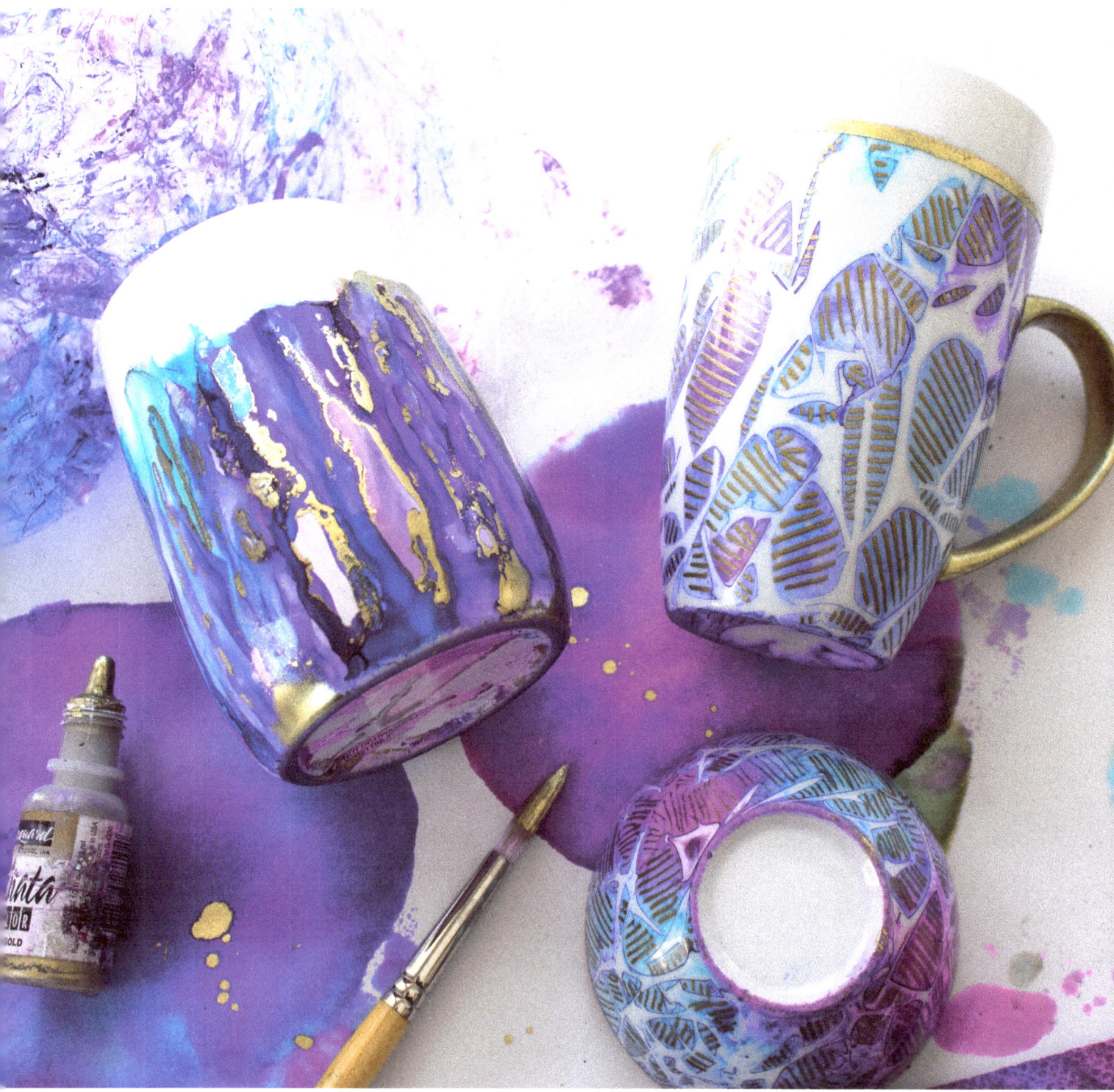

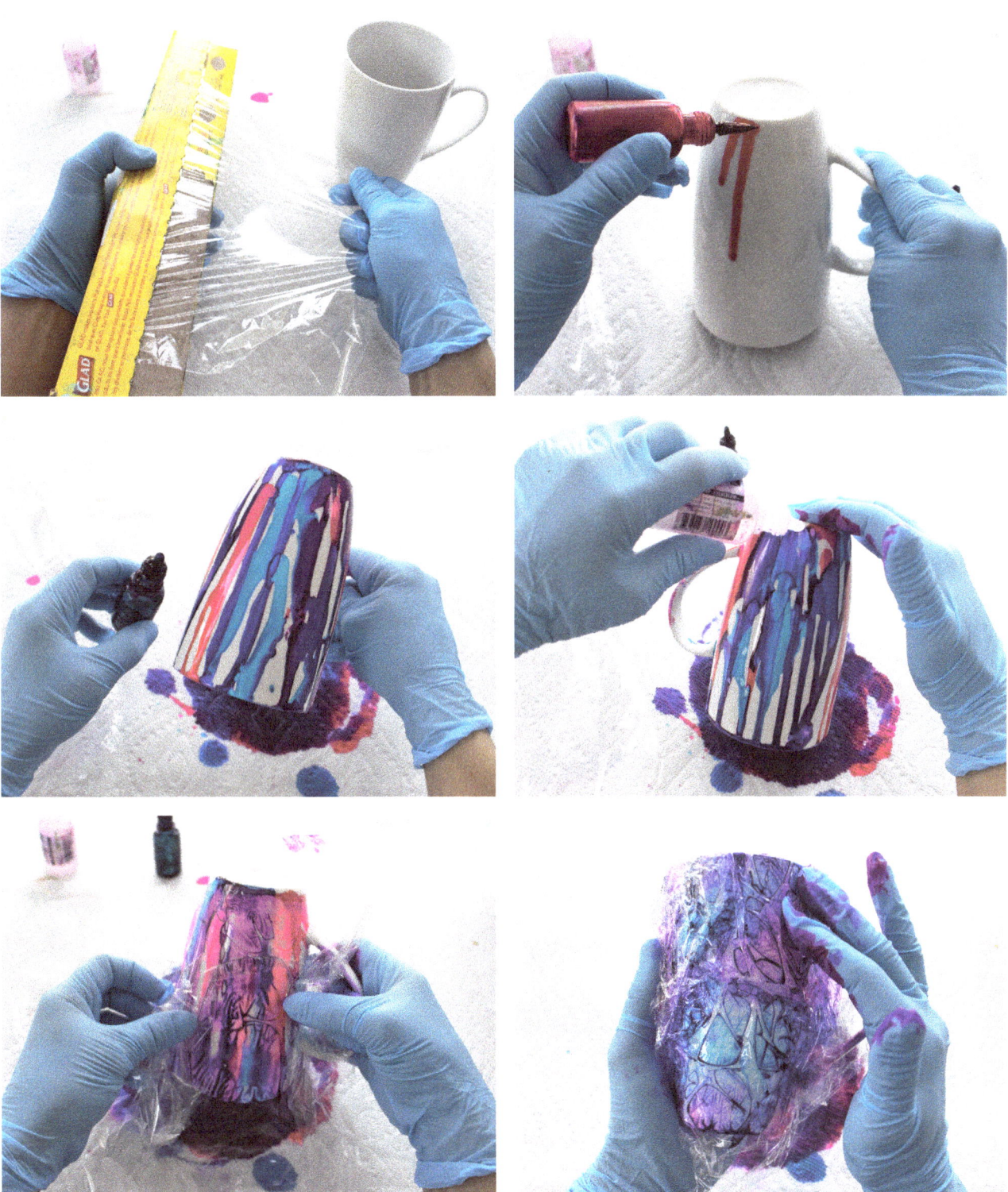

1. Tear off a piece of cling wrap and set aside. Begin with the same process as the previous ceramic vessel technique and drop several colors around the mug until covered. (Don't worry about smudges inside or on the top of the mug as it will be wiped clean in the final step.) Apply a generous amount of blending solution or alcohol around the mug. While still wet, quickly apply the cling wrap around the entire mug, mashing the cling wrap until the colors are blended nicely. Add more inks under the wrap if needed. Create mini pockets of air by pressing and moving the cling wrap until you see trapped alcohol form into intricate patterns. Leave the mug standing and do not touch for 24 hours.

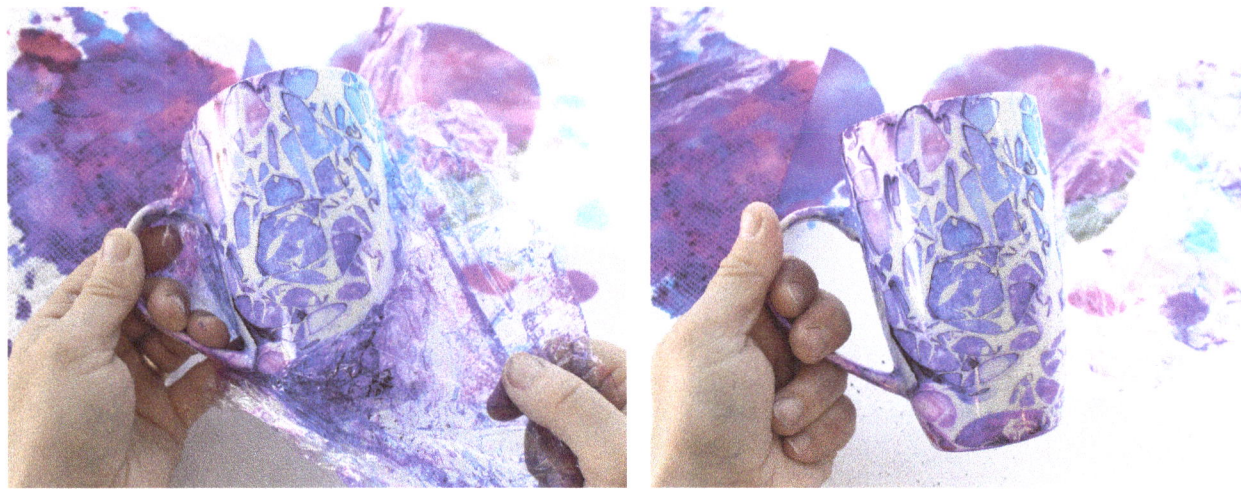

2. After 24 hours, remove the plastic wrap to reveal the dry ink detail. Do not attempt to remove the cling wrap earlier because the inks will not have dried and the effect will not be as dramatic.

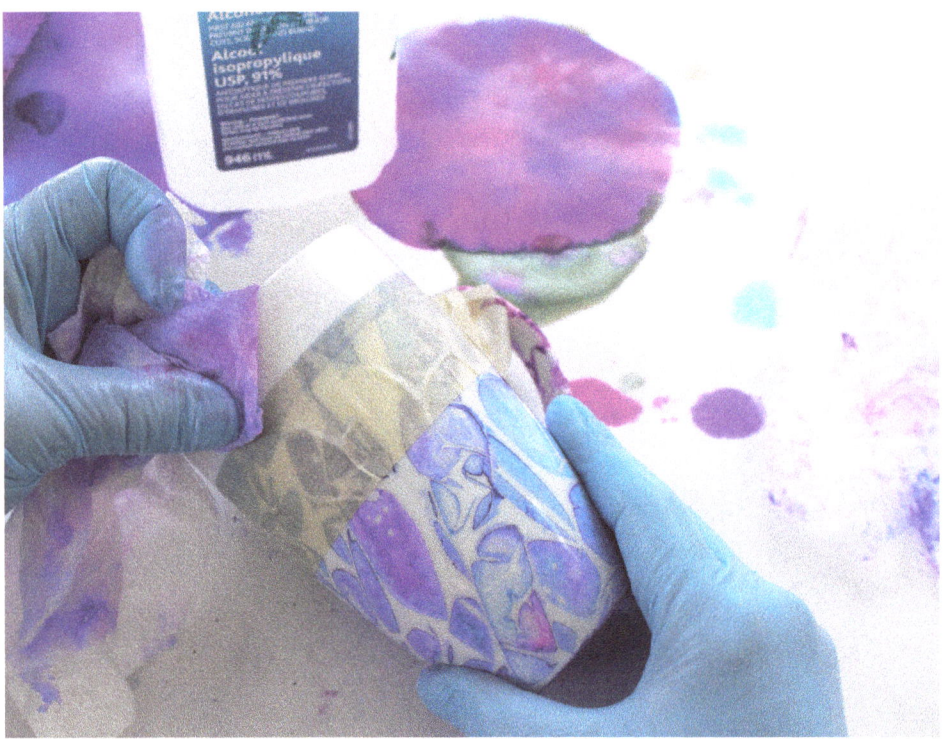

3. You will want to clean up the top area of the mug for a lip line and remove any smudging that may have occurred inside the mug. Tape off an even area around the mug to mask and allow for removal of ink above it. Wipe clean with alcohol until all traces of ink are gone.

TIPS

- Products used on these ceramics are not deemed food safe. When creating this technique on a mug, be sure to leave a gap at the top of the mug for health safety.

- Break up the gold into particles for a softer look. Add an alcohol drop immediately after dripping the metallic onto the ceramic vessel.

- For a softer look, use fewer inks and more blending solution or alcohol.

- Save your ink-dried paper towel and add alcohol to use as a swipe technique on Yupo paper.

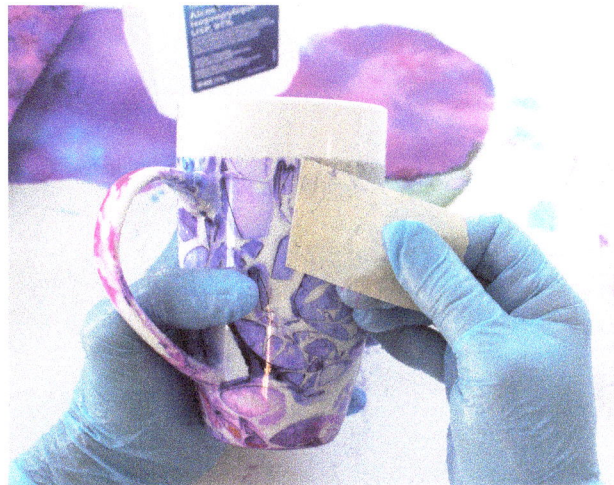

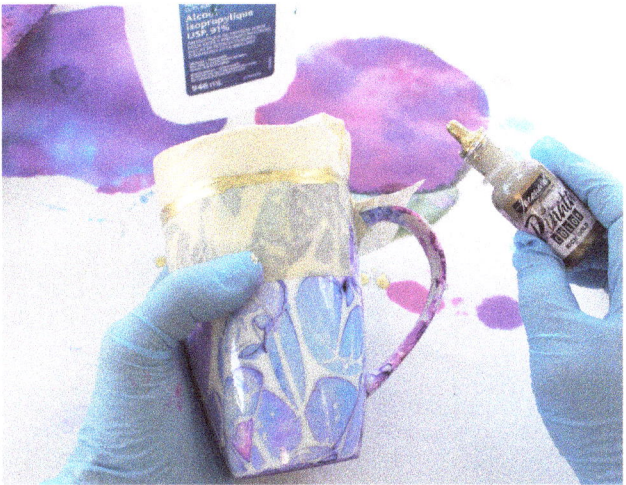

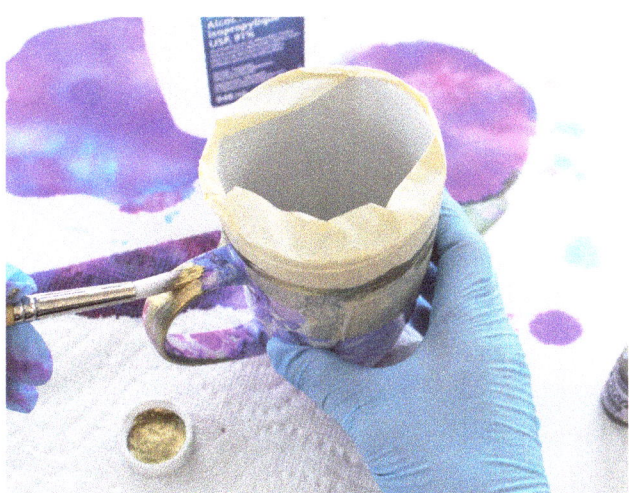

4. Remove the tape and remask a smaller, fine-width line around the top edge where the ink begins. Use a small brush to paint gold alcohol ink in the taped line. Paint the handle gold to match. Lastly, to add extra detail, create some lines with a gold marker in the negative space (white areas) or over the colored patches created by the plastic wrap.

5. To seal and protect your ceramics, spray with several coats of Krylon Kamar sealant. Allow each coat to dry before applying another. Additional UV protection should be considered too. Krylon or Golden UV Archival sprays work well. These items are not recommended for the dishwasher.

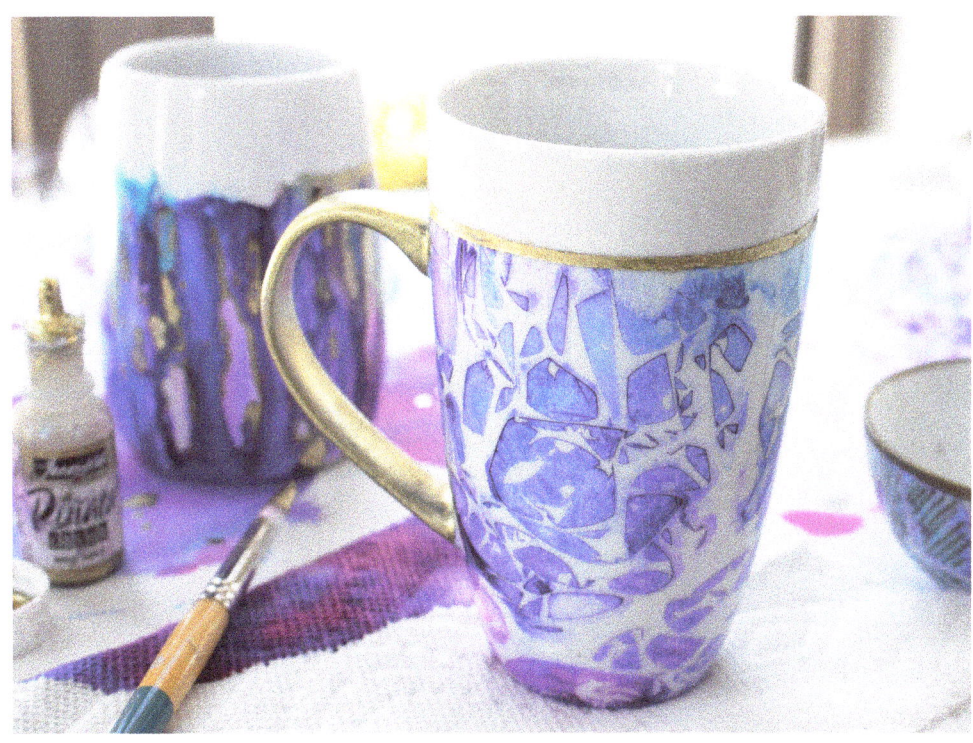

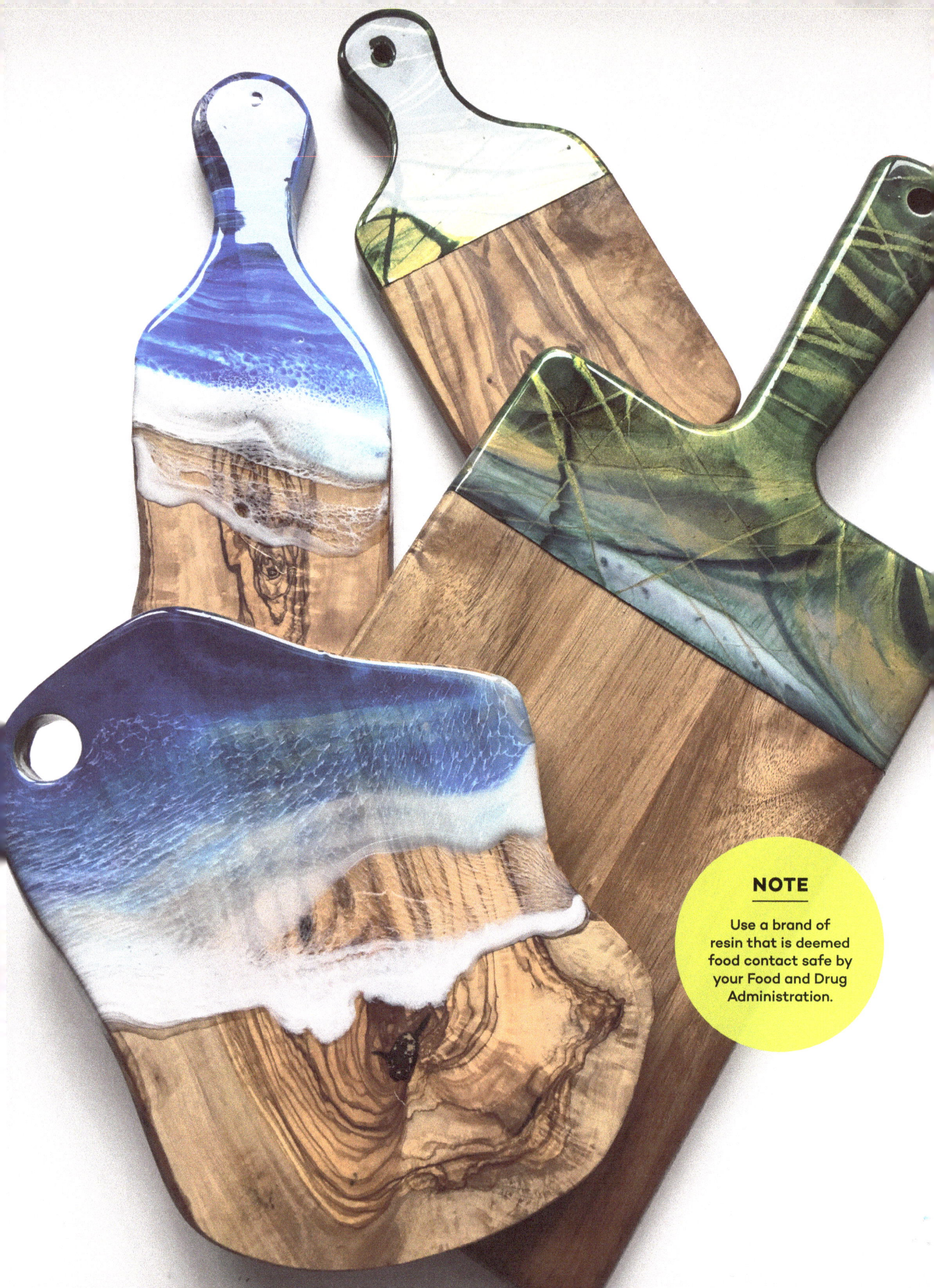

NOTE

Use a brand of resin that is deemed food contact safe by your Food and Drug Administration.

RESIN POUR CHEESEBOARDS IN TWO STYLES

I love how you can apply resin over just about anything! Wood is quickly becoming a popular choice. From raw wood tabletops to charcuterie boards, you can create some beautiful and functional resin art masterpieces. In this project, you will create two styles of cheeseboards—an abstract, straight-edge board and an ocean-style board with waves.

MATERIALS

black garbage bag or plastic tablecloth to protect surface area

painter's tape

wood board of choice

scissors

paint pen or brush (optional)

mixing containers

resin kit

individual flexible cups

mica powders

craft sticks

protective gloves

level

mini butane torch (optional)

container or box to cover artwork

X-Acto knife (optional)

medium-grit sandpaper

heat gun (for technique 2)

TIP

Applying resin over wood can sometimes produce off-gases, which can create micro bubbles in your finished project. When working with raw wood, it is best to first seal the area where you will be adding resin with a brush-on wood sealer.

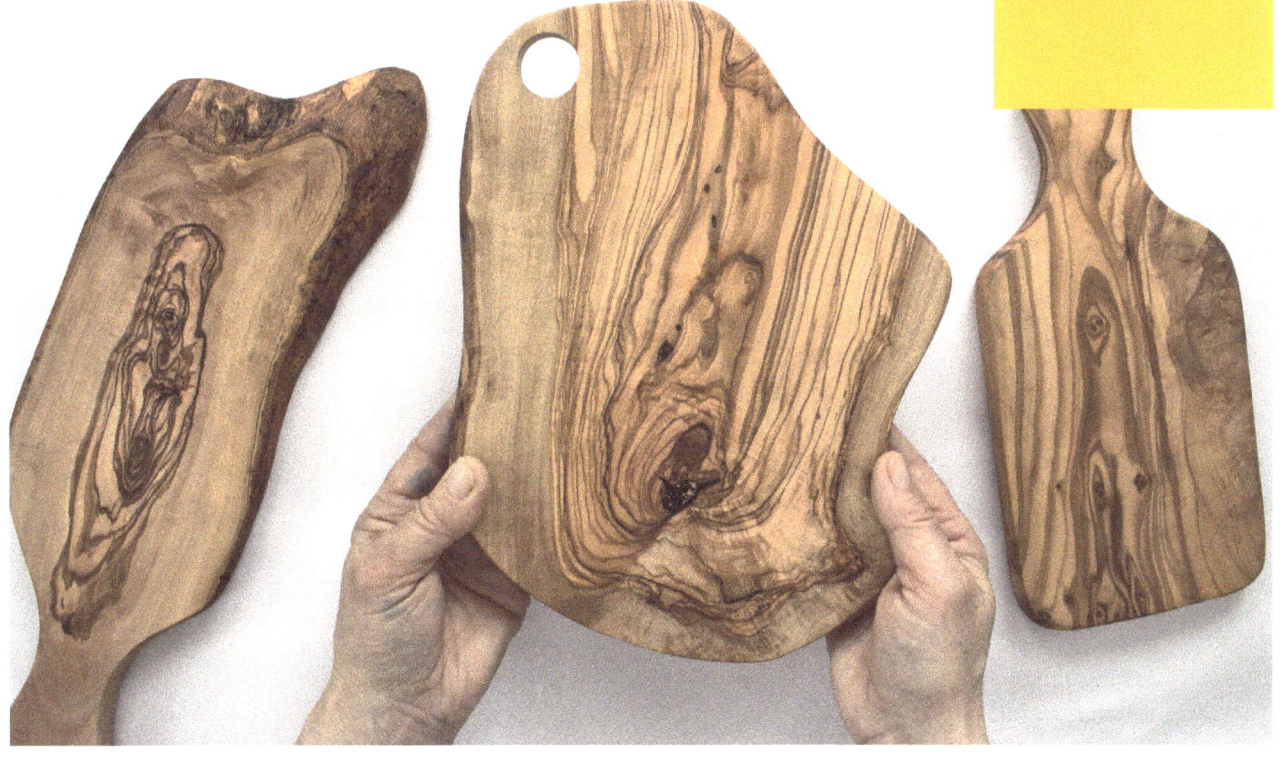

STRAIGHT-EDGE LINE BOARD

In this first project, you will create an abstract style with colored lines and a straight edge. You will tint your resin with two mica powder colors— a gold metallic and an opaque color of your choice.

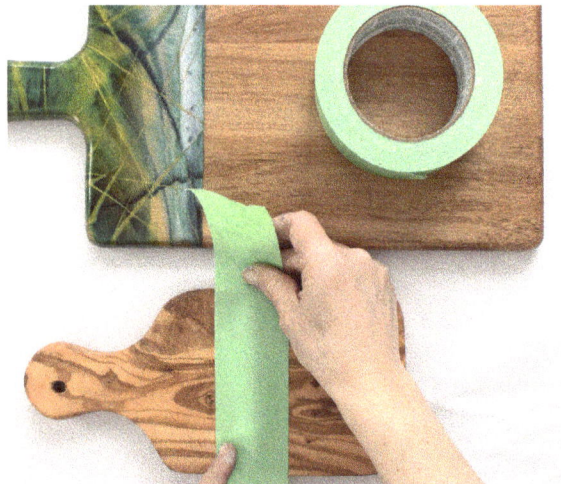
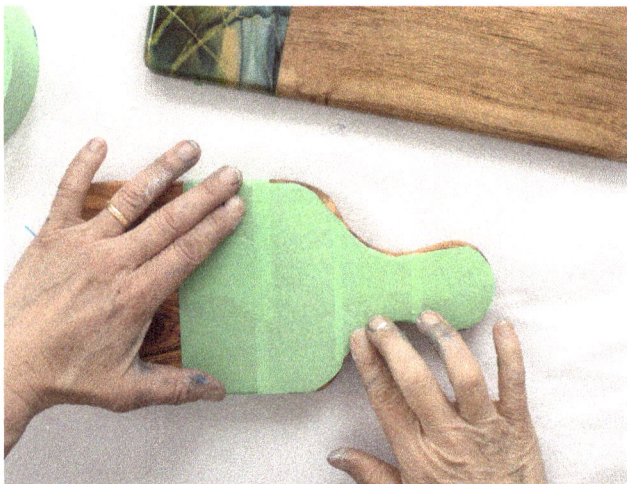

1. Begin by covering your work surface with a black garbage bag or plastic tablecloth to protect the area from dripping resin. Then use painter's tape to add a straight line across the top of your board. Tape over and around the backside too. Next, cover the underside of the board to protect the area from resin runoff and drips. Cut the tape to the underside edge of the wood.

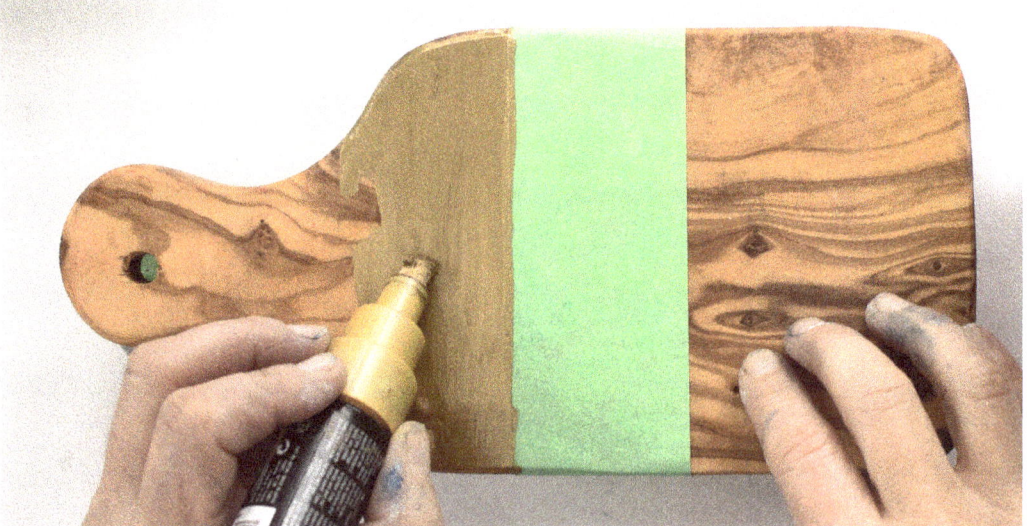

2. You may wish to paint the surface area using a paint pen or brush before pouring your first layer of resin. This is optional, but it will create a solid background and produce less of a transparent effect, allowing you to work with just a single coat of resin instead of multiple layers. It also helps if you miss coating a side area with resin, as the paint will appear to be part of the look. Paint the surface with a color that is similar to the resin colors you are using.

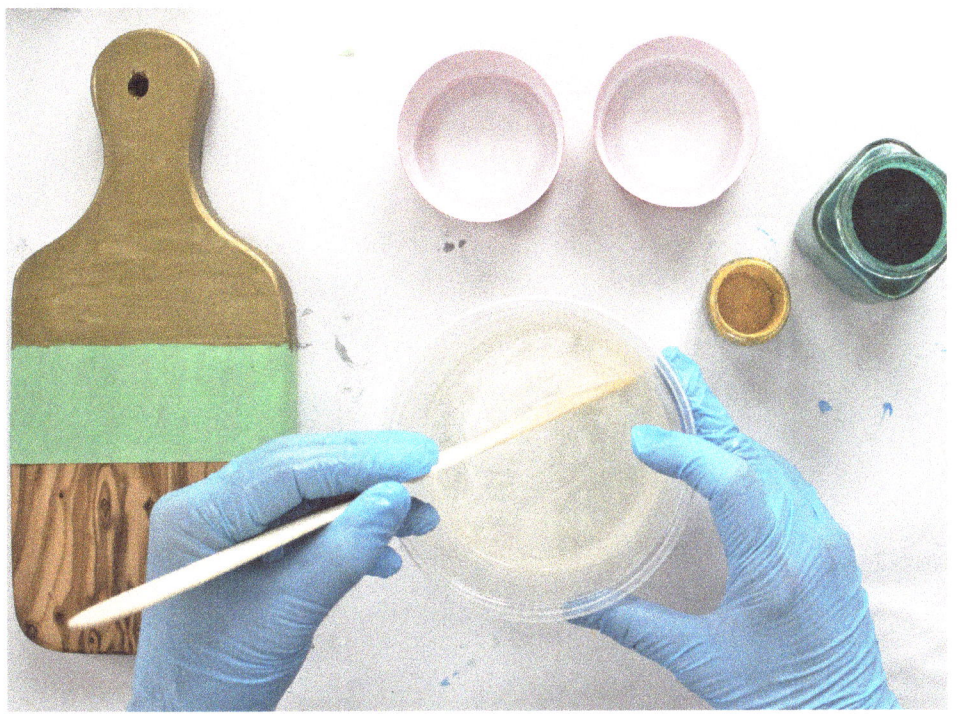

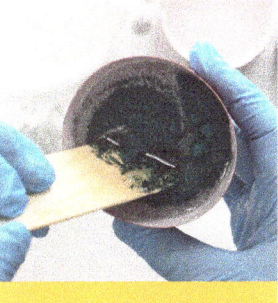

TIP

Mica powders are highly pigmented, so only a small amount is needed. Start with ¼ teaspoon (1 g) or less and mix well to prevent pigment particles from getting trapped in the cured resin. Continuing to add powder will create a more opaque color.

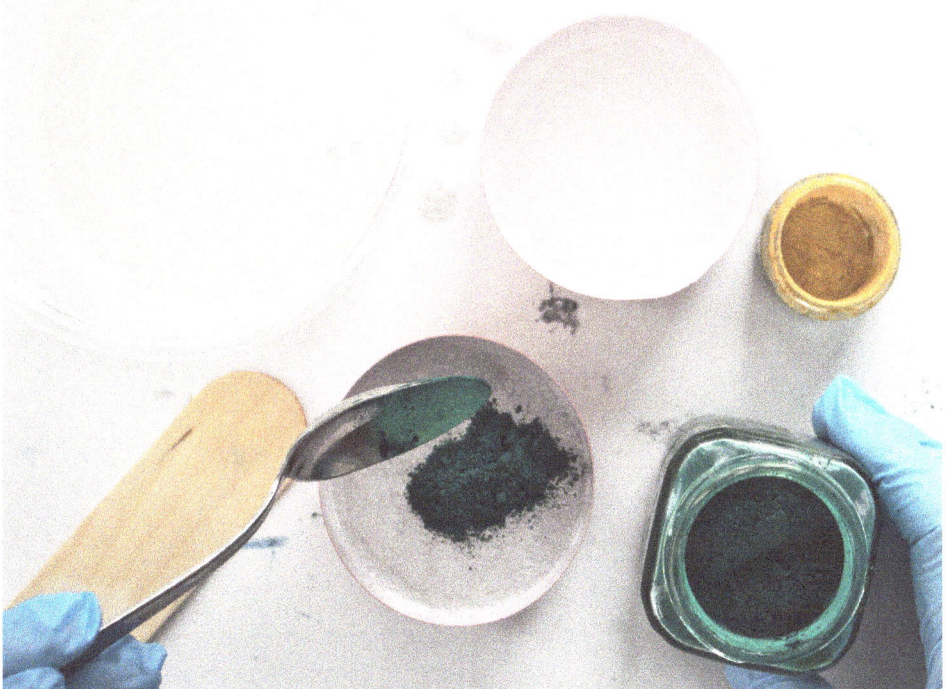

3. Next, measure and prepare your resin according to the product instructions and pour into several individual flexible cups (trimmed-down red plastic picnic cups work great). Add a small amount of mica powder to each cup and stir with a craft stick.

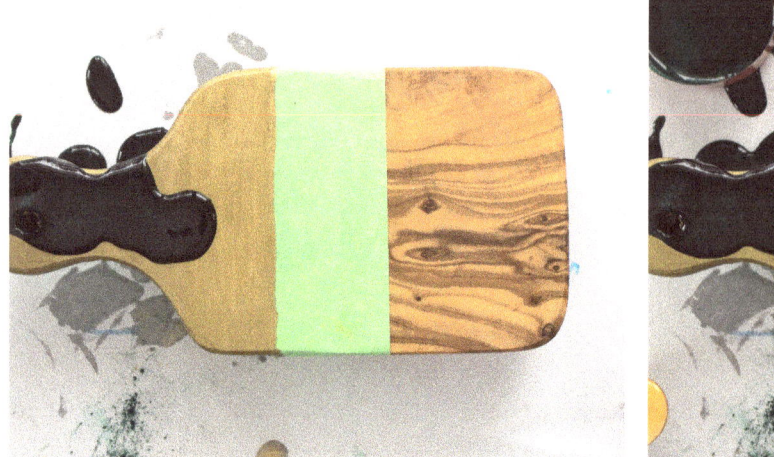 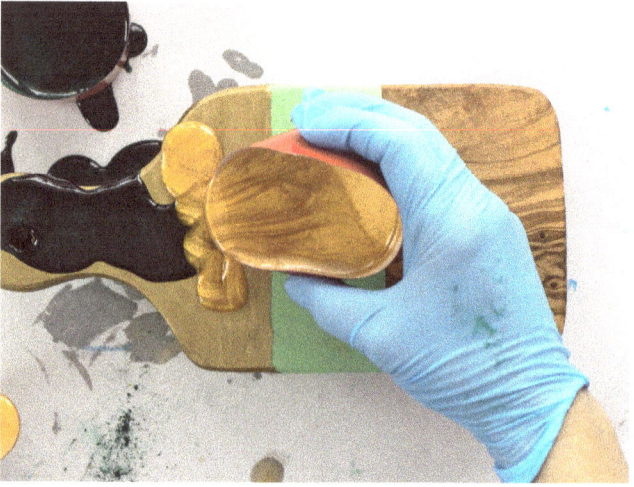

4. Working on your protected surface with gloves on, place your board slightly above the work area and make sure it is level. Mini paper/plastic cups work well and can be discarded during cleanup. Begin by pouring some of the first tinted mixture from the top of the handle to halfway down the board. Then continue with the second color.

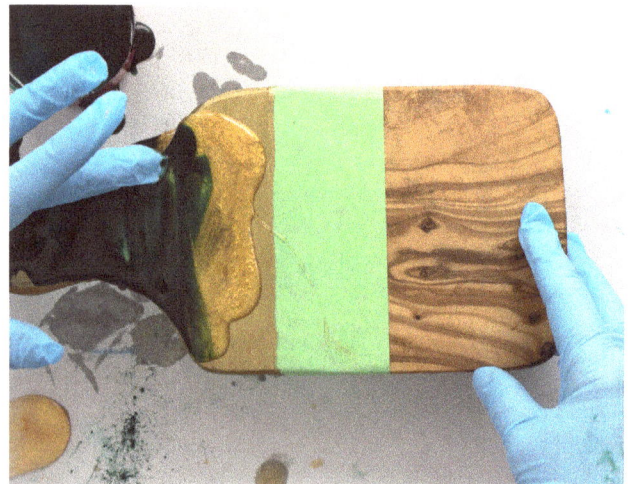 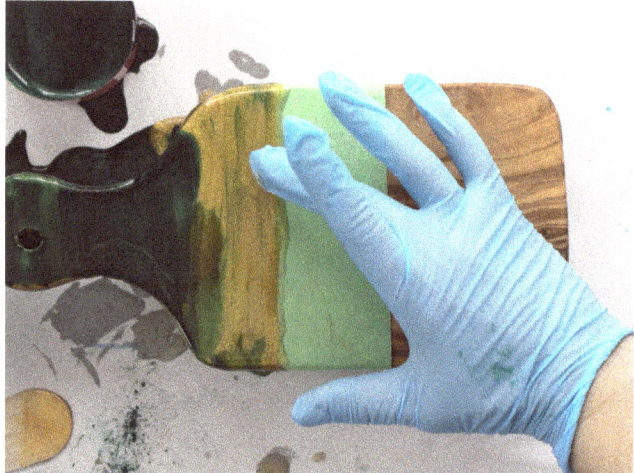

5. With one of your glove hands, gently blend the colors together until you get a nice, smooth, subtle pattern.

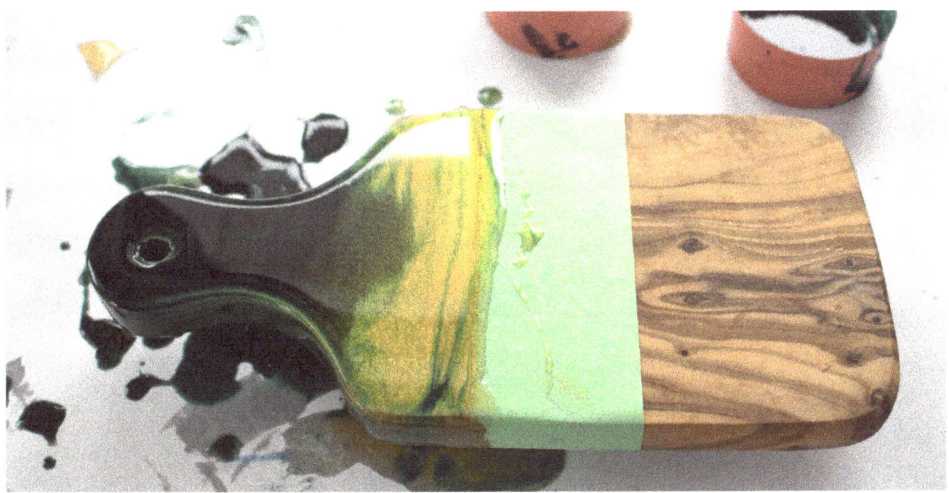

TIP

Lines created in resin immediately after mixing will not hold the pattern and will fade or shift during curing. Allowing a small amount in a container to sit and thicken for a bit first will improve your chances of maintaining details.

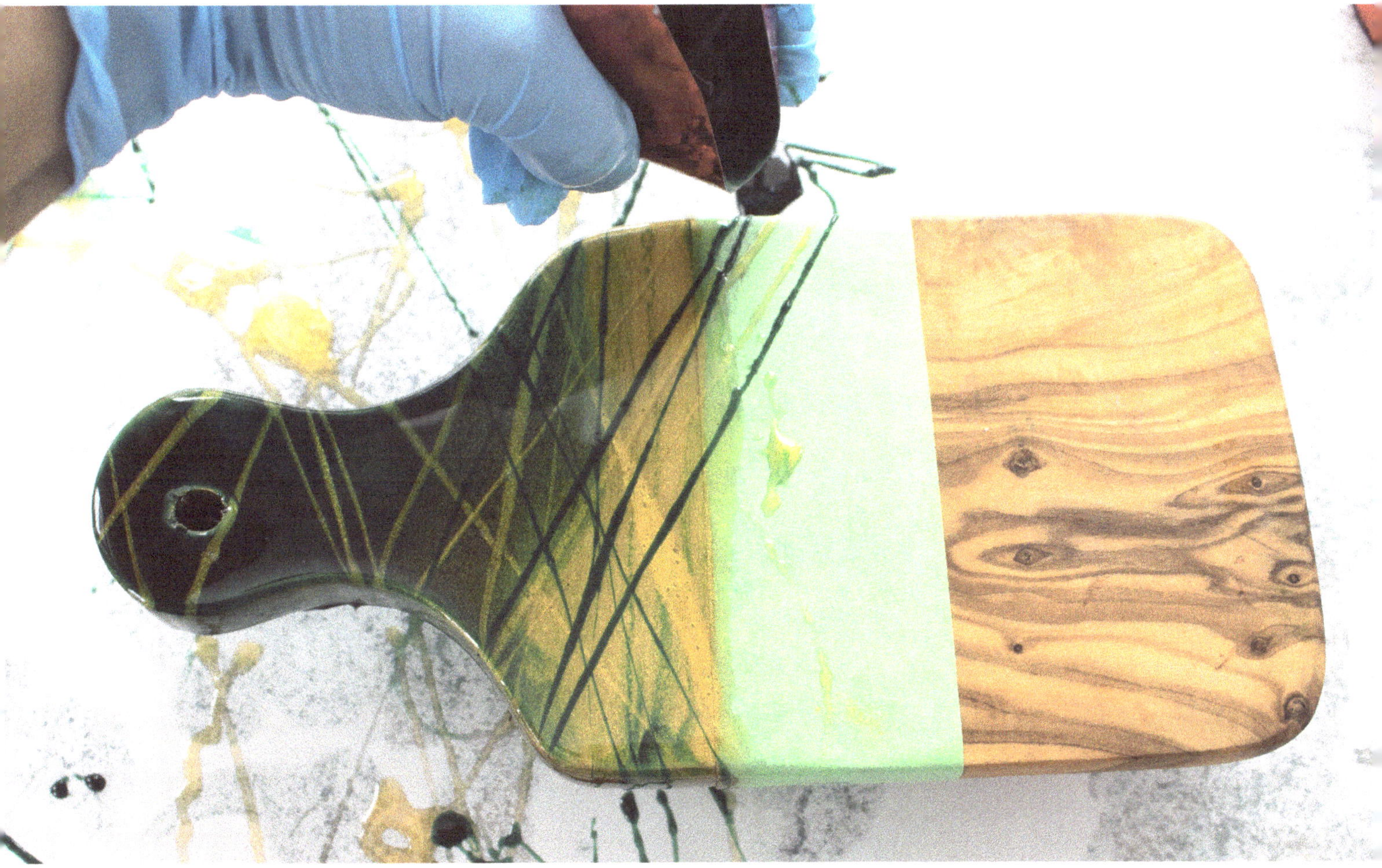

6. Allow your resin layer to rest for about an hour or until the leftover mixture in your cups are the consistency of a thick syrup that spills very slowly when poured out of the container. This will produce a more defined design and will hold its shape when curing.

7. Squeeze the cup to form a spout that will create a line when you pour it across your board. Alternate between the colors until you are happy with the design. You can torch the surface lightly to remove any bubbles before allowing to cure (see project 2 for how to use the butane torch).

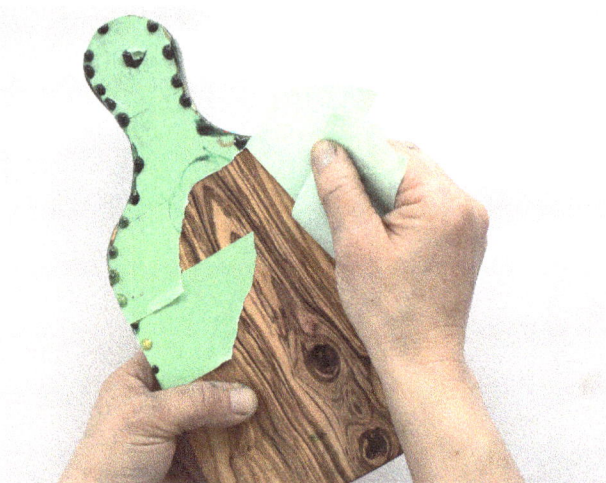

8. Cover your project with a sizable container to prevent dust from settling into your resin and let cure for at least 12 hours. Remove and peel the tape from the back of the board.

TIP

If you are having trouble removing some of the tape, carefully warm it with a heat gun to soften the resin drips. This will make it easier to lift the tape.

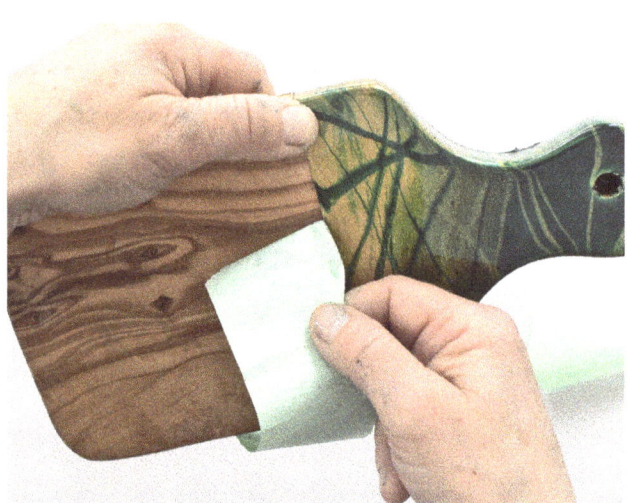

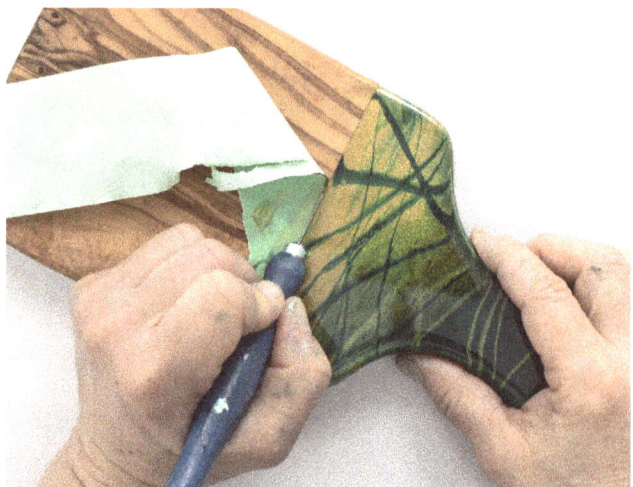

9. When pulling off the tape from the front of the board, do it slowly to avoid ripping or tearing the resin, especially because it is not yet fully cured. You can also lightly score the tape line by running an X-Acto knife across the board to get a nice, crisp finish.

NOTE

Resin can withstand dishwasher use, but it is not recommended to wash wood in a dishwasher. Hand wash and wipe your cheeseboard.

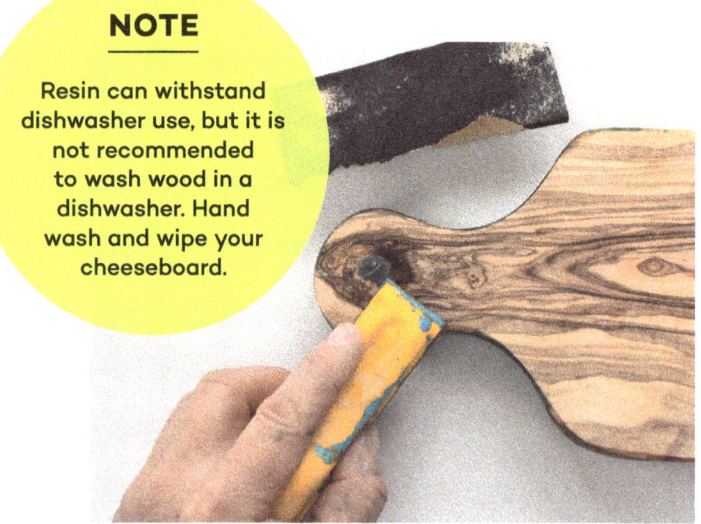

10. After all the tape has been removed, check the back of the board for areas of resin that may remain. Lightly sand with medium-grit sandpaper. If there's a lot of dried resin, use an electric hand sander to get an even, smooth finish.

11. The board will fully cure over the next 24 hours and your resin will become completely hard. You may wish to season your board with some walnut oil. You're now ready to enjoy your cheeseboard.

OCEAN BOARD

Ocean-style art is becoming increasingly popular, and when the right tools are used, realistic beach and ocean waves can be produced. In order to create the laced, wavy effect of the ocean, you MUST use a white pigment resin tint and a heat gun.

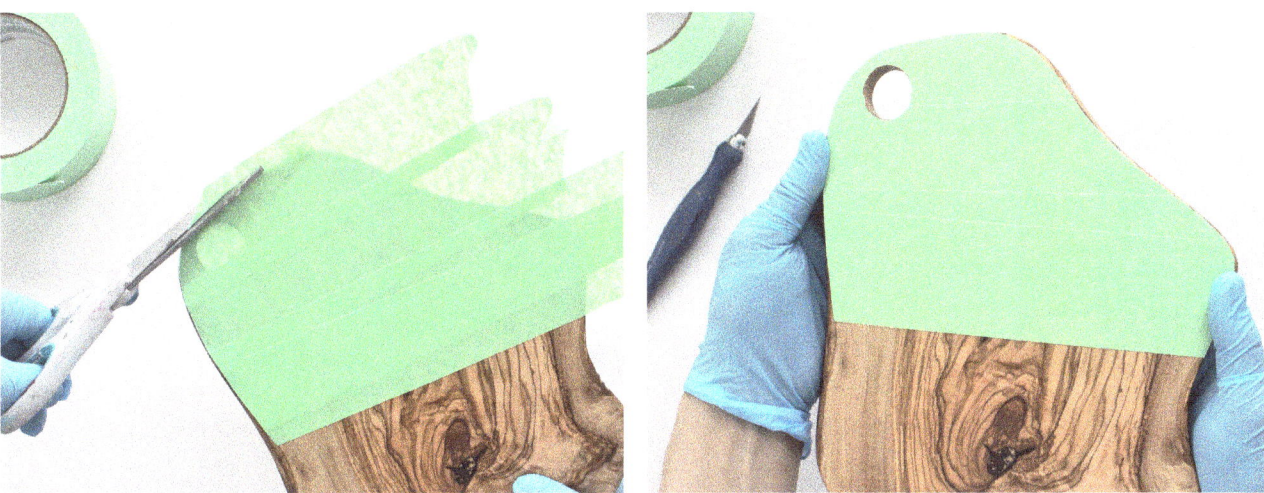

1. In this project you do not need to tape the front of the board because the resin will be flowing across the top. Only the back of the board needs to be protected. Tape and cut to the back edge of the board. Cut out the circle area of your cheeseboard if it has one.

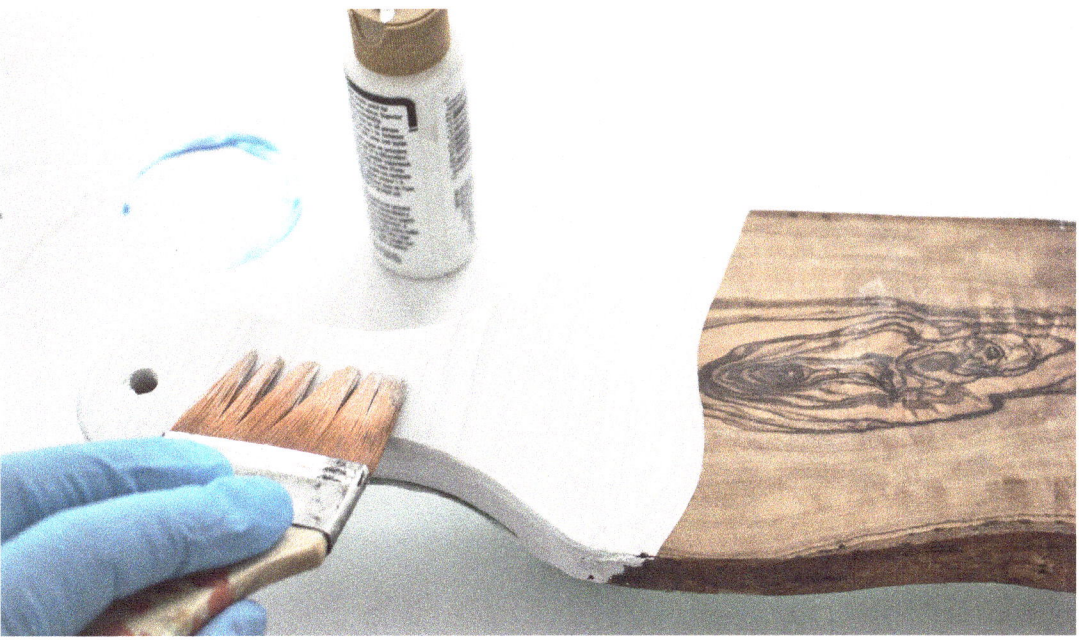

2. You will be creating several layers of transparent resin for this style to produce a realistic effect. However, you may choose to also paint the surface of the wood, as in the first techinque, to create a less transparent first layer. Experiment with the process until you find the results you like best.

WHY AND WHEN TO USE A HEAT GUN VS. A BUTANE TORCH

A butane torch has a flame. A heat gun blows hot air.

One of the most frequently asked questions I get when it comes to resin is which heat tool to use. I use both a butane torch and a heat gun for resin. Each is used for different purposes. If you are adding a clear coat of resin for a finished, glass-like look, you should opt for a butane torch. The flame from a butane torch eliminates the bubbles from the surface and produces a glass-like finish without moving the resin. If you want to mix colors and push resin around to create texture and depth, use a heat tool. A heat tool blows hot air onto the surface, allowing you to manipulate the resin and create different effects, such as waves on a beach. (A hair dryer will not have the same effect because it is not hot enough and blows too much air.)

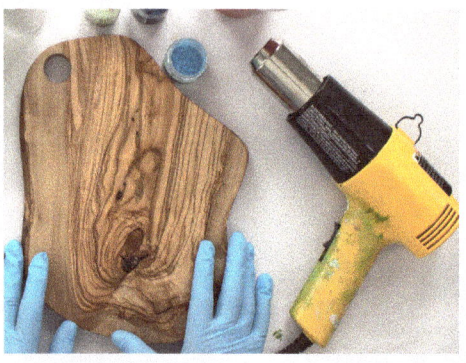

A heat gun will push around your resin to create texture and movement.

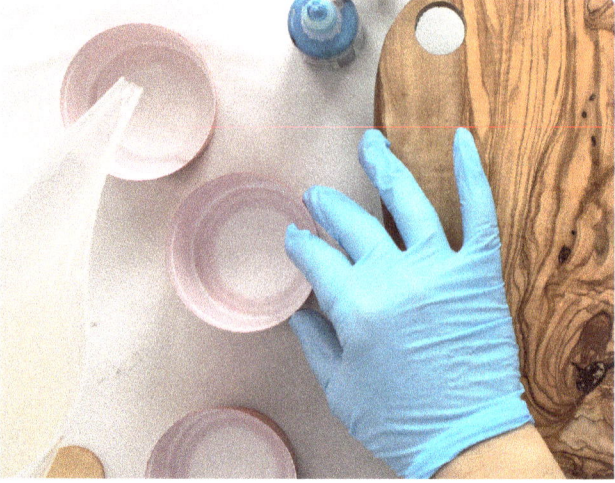

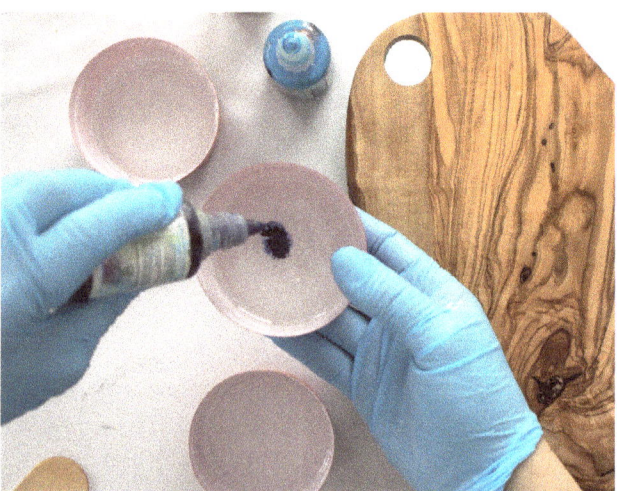

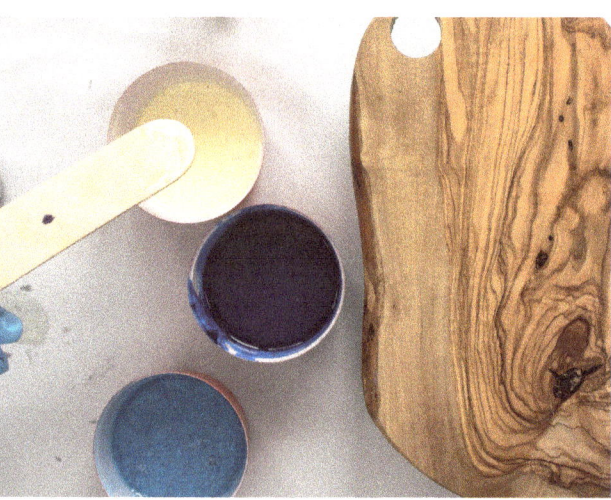

3. Wearing gloves, mix your resin and pour into individual cups. This time you will be adding resin tints in several different blues and a white. Add a few drops of tint to each cup of resin and mix thoroughly. Have some clear, uncolored resin in a container ready to use as well.

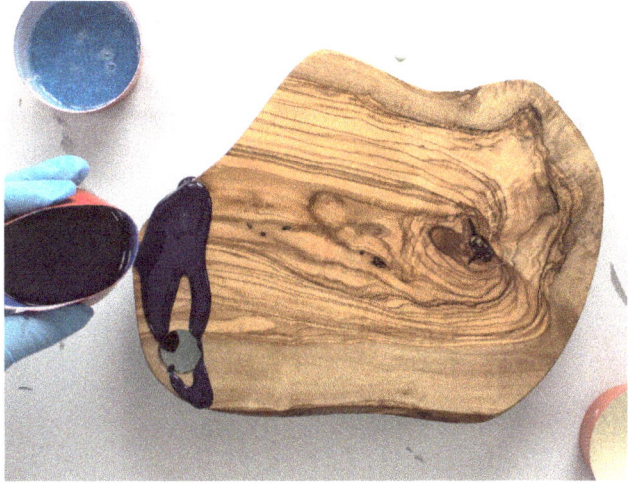

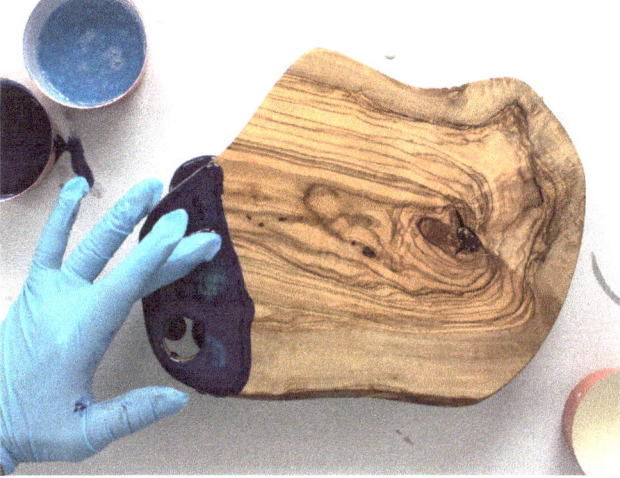

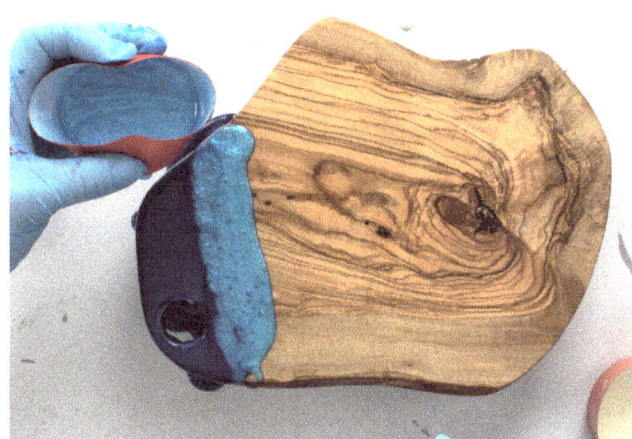

4. Begin by pouring the darker blue over the top end of the board, followed by a lighter blue. You will want to gradually create a gradient that represents the change in ocean color. You can blend a little with your gloved hand too.

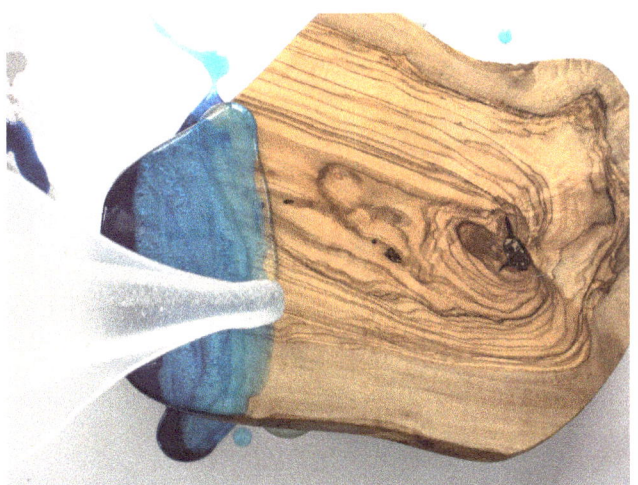

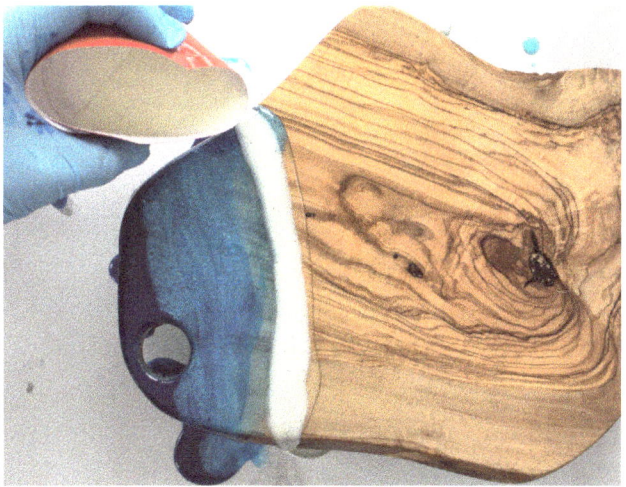

5. To create a realistic wave effect, pour a layer of clear resin that slightly overlaps the last color before you add a line of white. Use the heat gun to push the white color around, especially when creating the second layer. This technique produces the most realistic lacing effect of ocean waves.

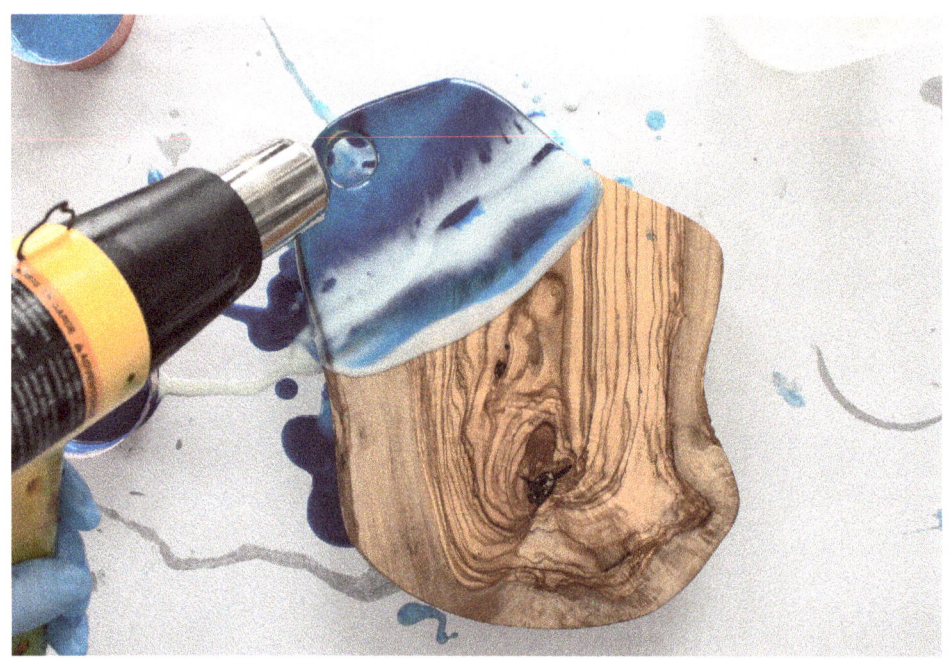

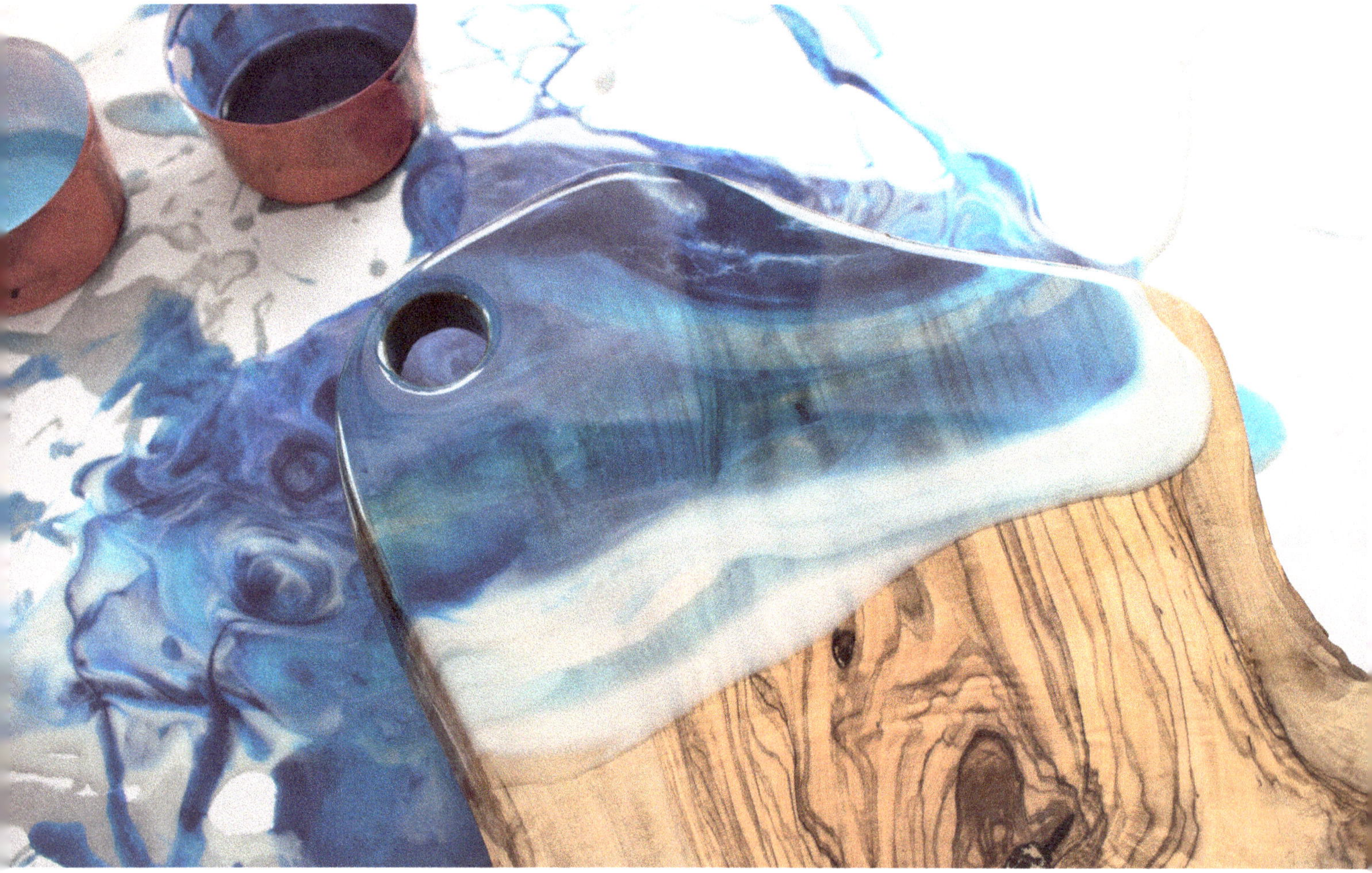

6. Use your heat gun to manipulate and push the colors around. This is your base layer, which is the most translucent layer. Don't worry too much if you don't get the exact look you want because you will be repeating the process with a second layer once the first has cured. Cover your board with a container and leave for 24 hours, or until fully cured.

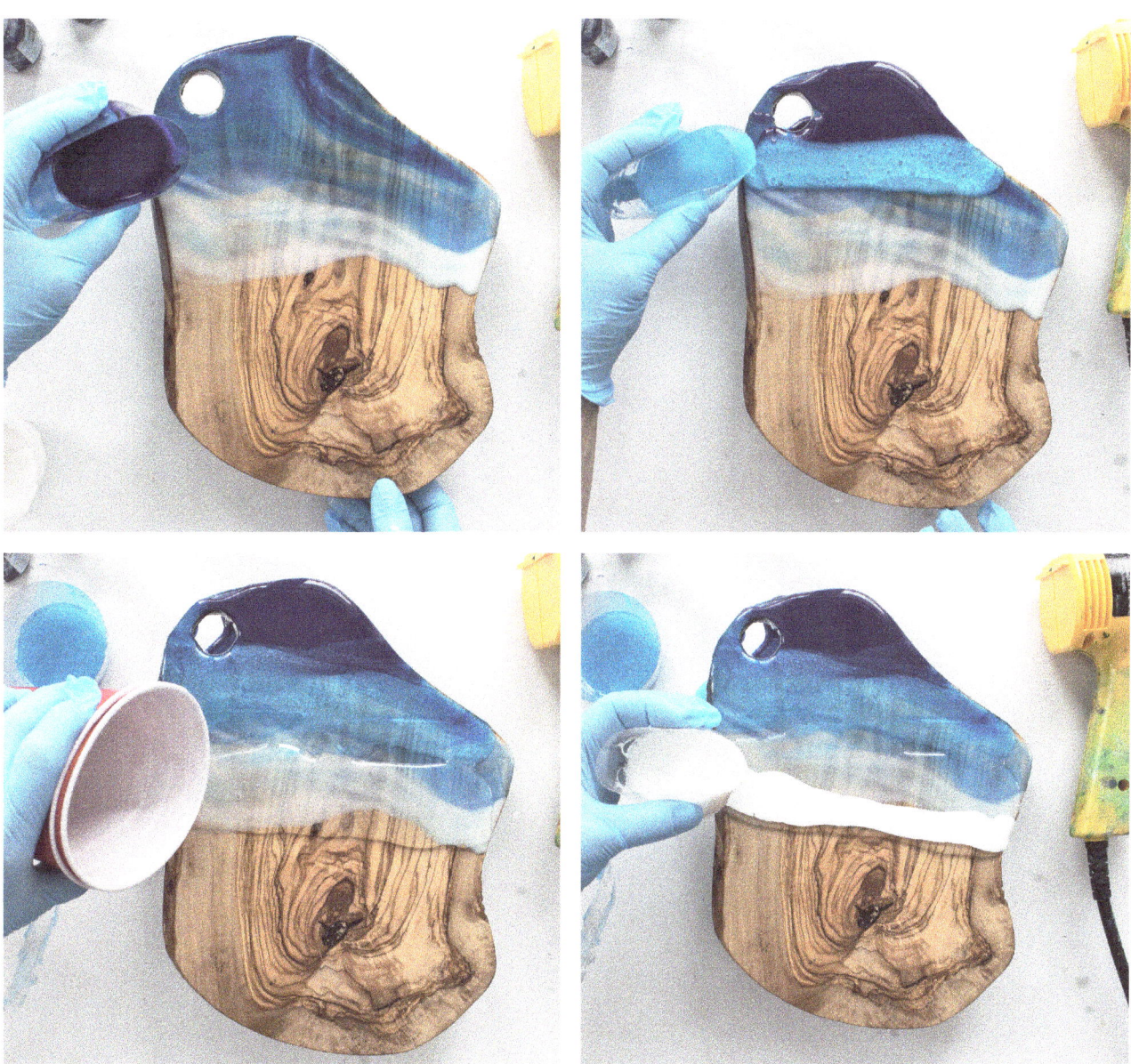

7. You will now repeat the same process on your now hardened first layer. More detailing and a realistic look can be achieved in your second layer. Mix up a new batch of resin and repeat with the same tint colors. Begin pouring as you did the first layer.

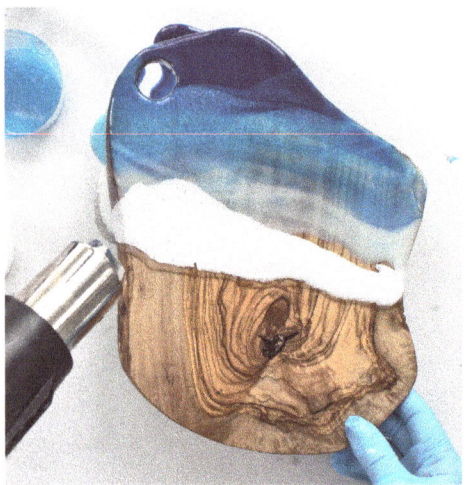
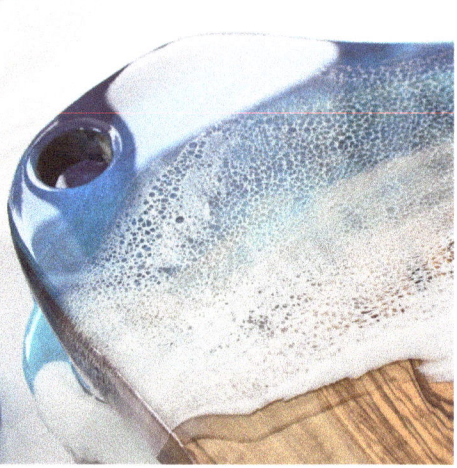

TIP

Experiment with different brands of white resin tint. Some yield whiter results and produce more lacing effects depending on the product's ingredients.

8. Once again, use your heat gun to start pushing the resin around until you see some separation and lacing in the white tint. This will happen quite quickly and once you've achieved the desired look, leave it alone. It may shift or change slightly, but it should remain close to what you have now.

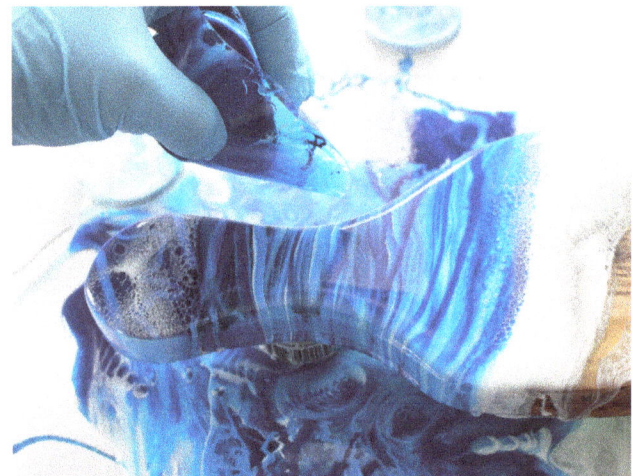
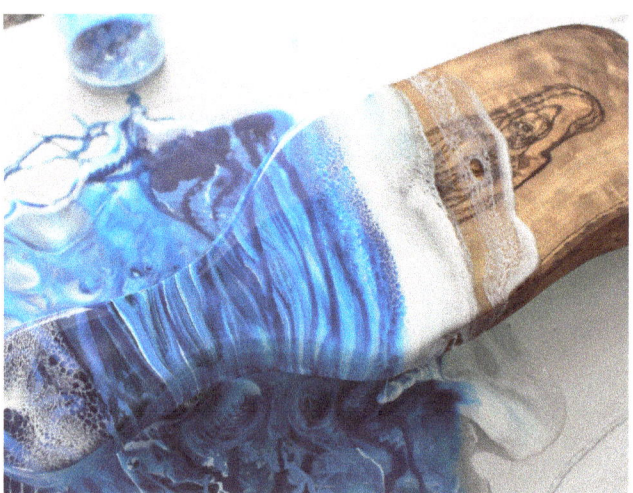
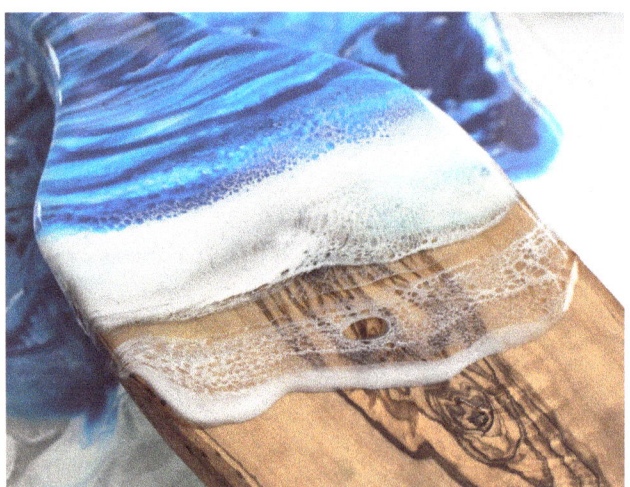

9. If you want more wavy line detail in the ocean, leave the colored resin in your cup to thicken, as you did in the first project. Then squeeze the cup to form a spout and pour lines onto the board.

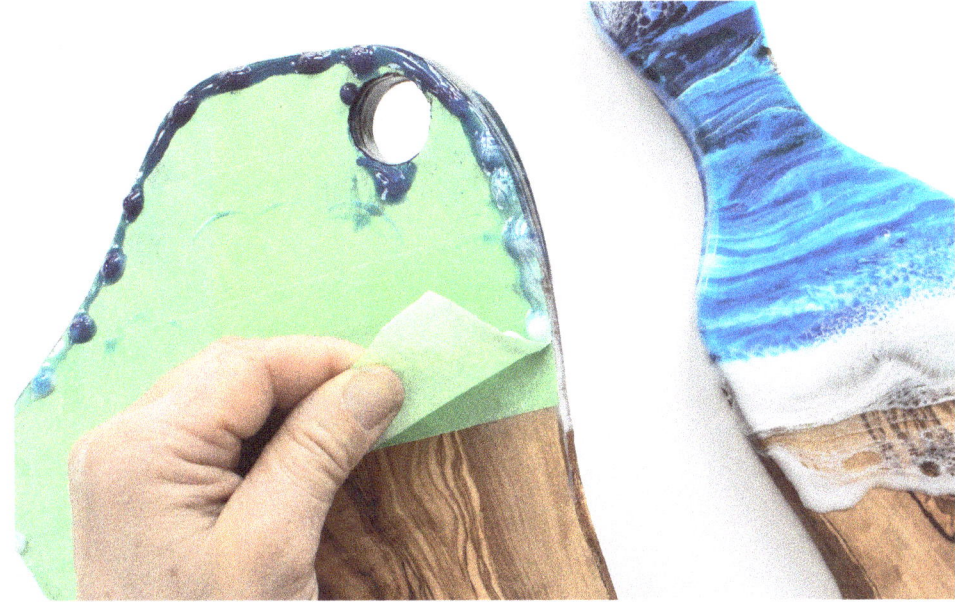

10. Cover your board with a container while curing to prevent dust particles from settling into the resin. Leave for at least 12 hours before removing the tape. Remove the tape from the back of the board and sand if necessary, as with technique 1.

Your ocean board will harden to a full cure over the next several days, at which point you can begin using and enjoying your modern piece of functional art!

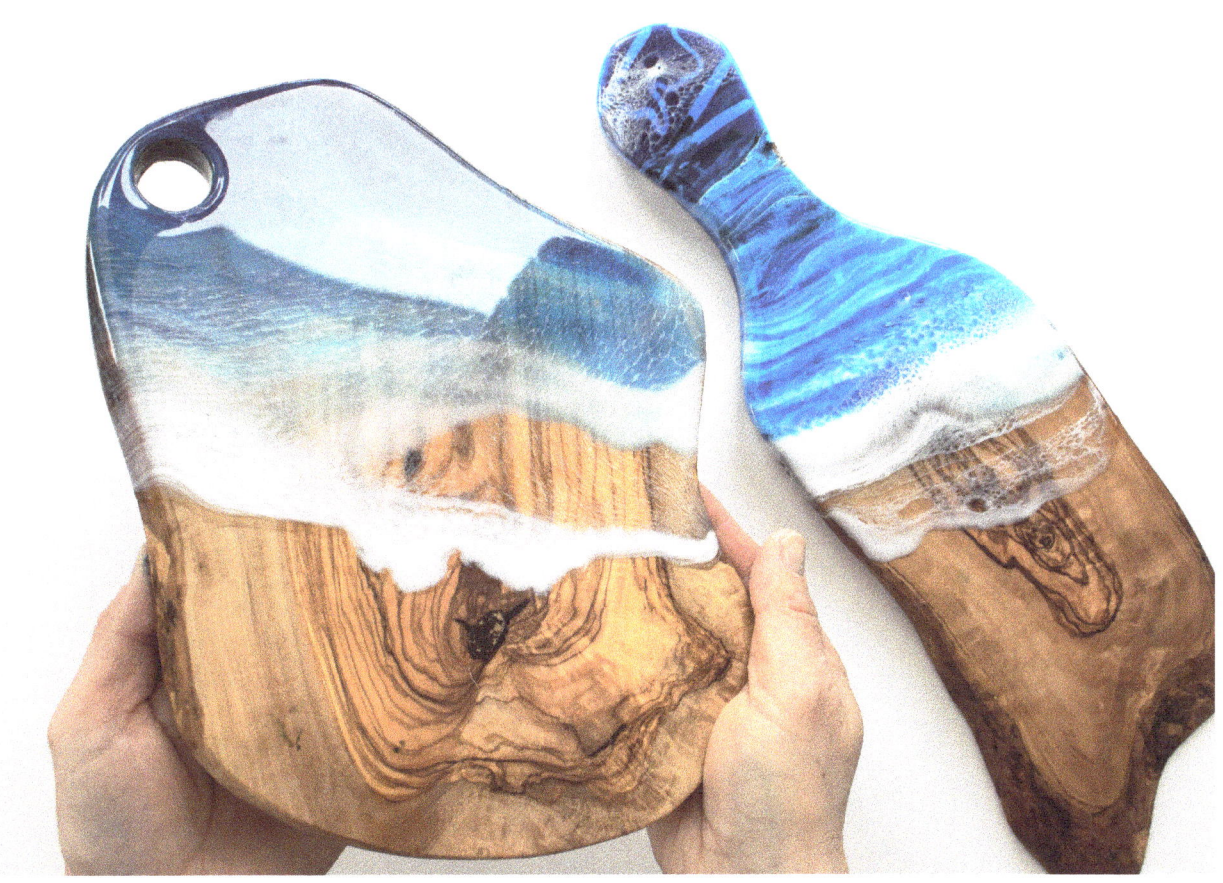

PROJECT 9
RESIN VASE

Using a resin with a longer working time allows you to be creative and make all sorts of fun crafts, including a mold-free vase that you can use for potted plants or fresh flowers. You can make any size, color, or pattern. And when fully cured, they will hold water!

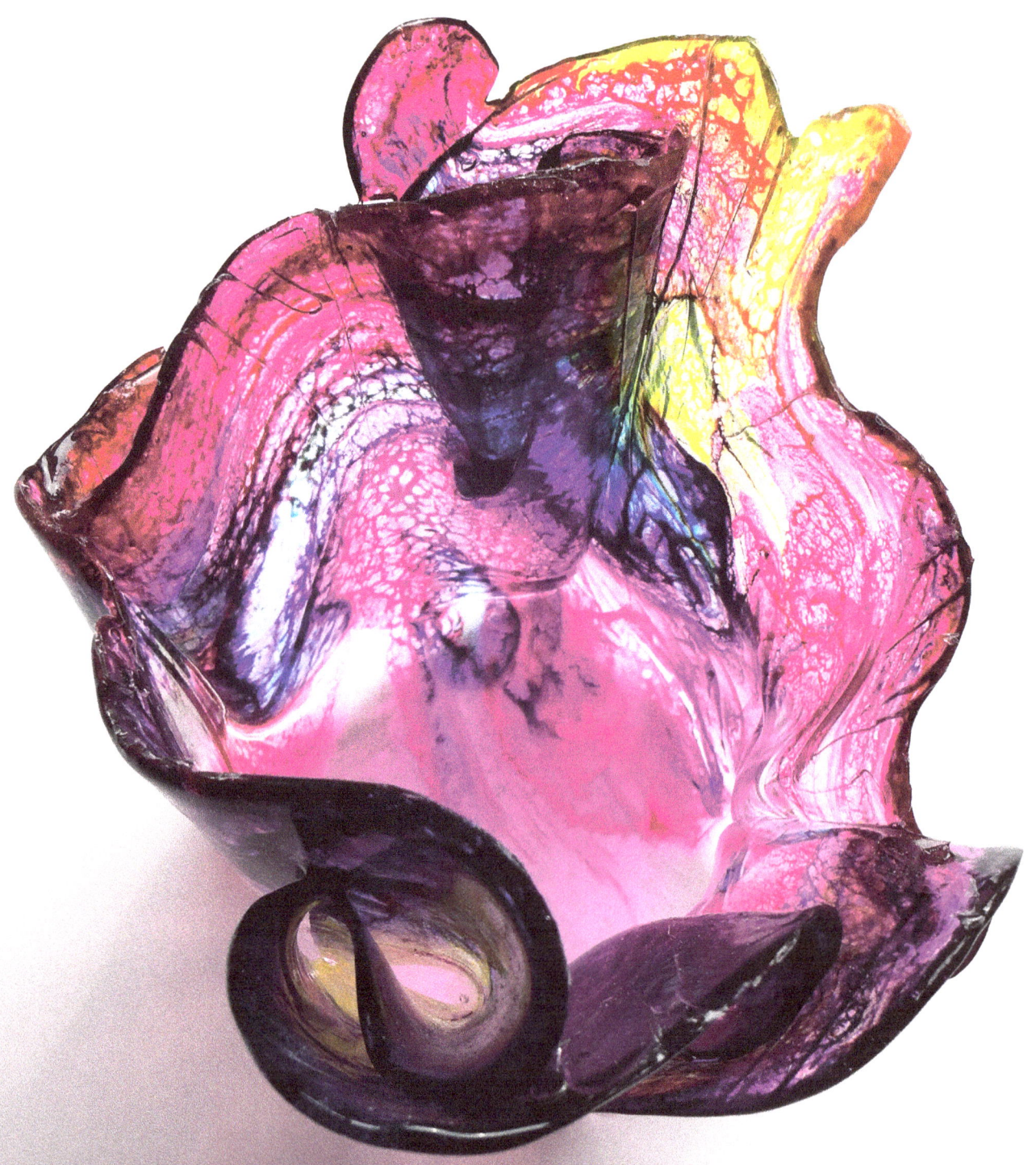

MATERIALS

black garbage bag or tablecloth

level

thick plastic liner such as a shower curtain

tape (optional)

tray or board (optional)

protective gloves

resin kit

mixing and measuring containers

craft sticks

large plastic vase or container

butane torch

alcohol inks

toothpicks

silicone oil

rubber bands

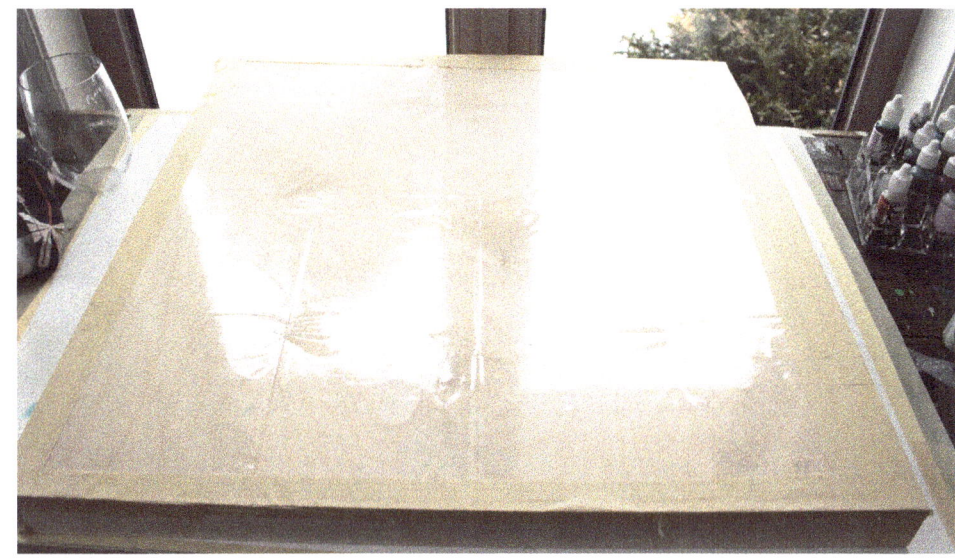

1. Begin by prepping your surface, making sure it is level and covering it with a black garbage bag or tablecloth. Arrange your plastic liner material on the surface, making sure it is smooth, flat, and taped down. You will be pouring your resin directly onto the plastic, so this step is important. Creases in your surface will appear in your cured resin.

You may choose to tape your material onto a tray or board so you can move it if needed.

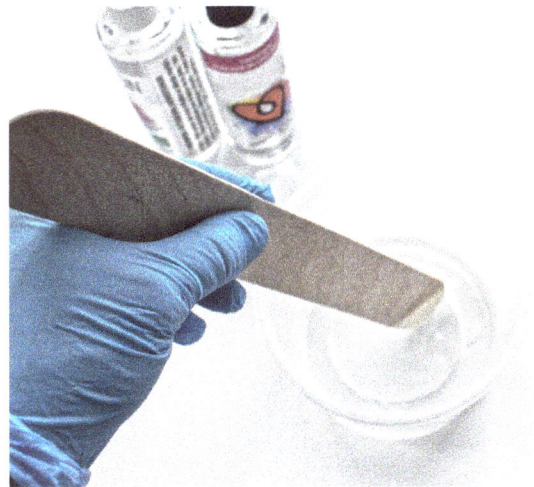

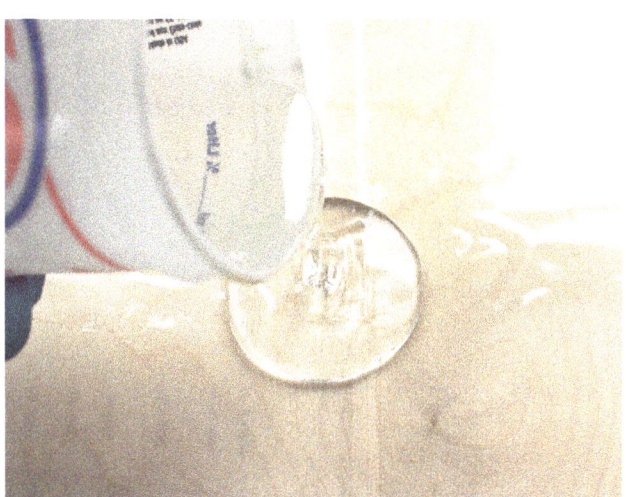

2. Wearing protective gloves, mix your resin with a craft stick and allow it to rest and slightly thicken before pouring onto your surface. This will produce a thicker vase and less spreading will occur on the surface. Pouring the resin immediately after mixing will create a thinner vase. Timing is dependent on the brand of resin you use as well as your level of experience working with resin. Using resin consistently in your projects will make you aware of working times, and better allow you to manipulate results.

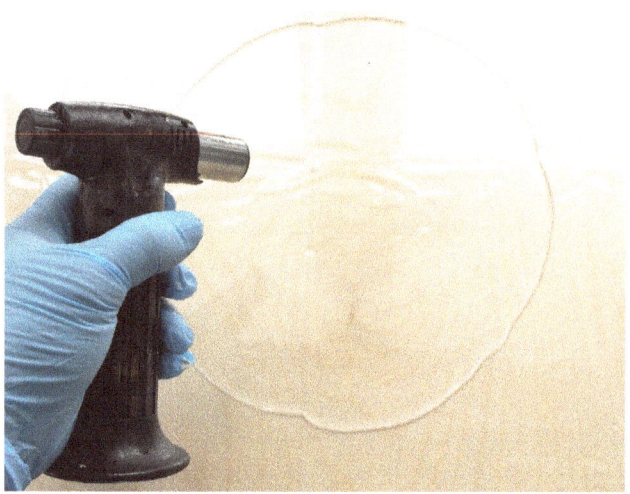 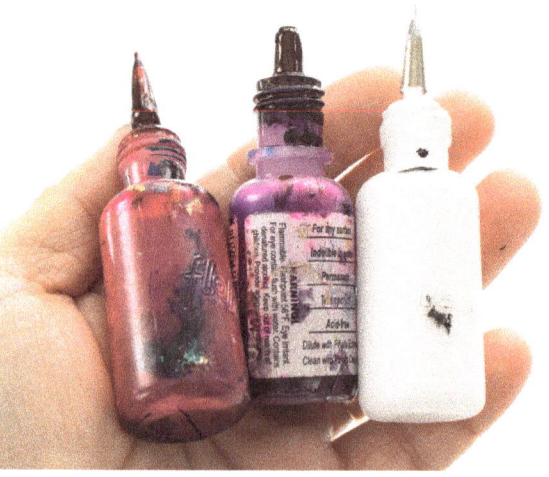

3. Pour enough resin onto the surface to allow you to cover the plastic vase or container that you are using as a form. Lightly torch the surface with a butane torch to eliminate surface bubbles.

4. Select several alcohol ink colors and apply several drops over the resin before using a toothpick to create a pattern. Remember that allowing your resin to rest before doing this step will result in a better pattern.

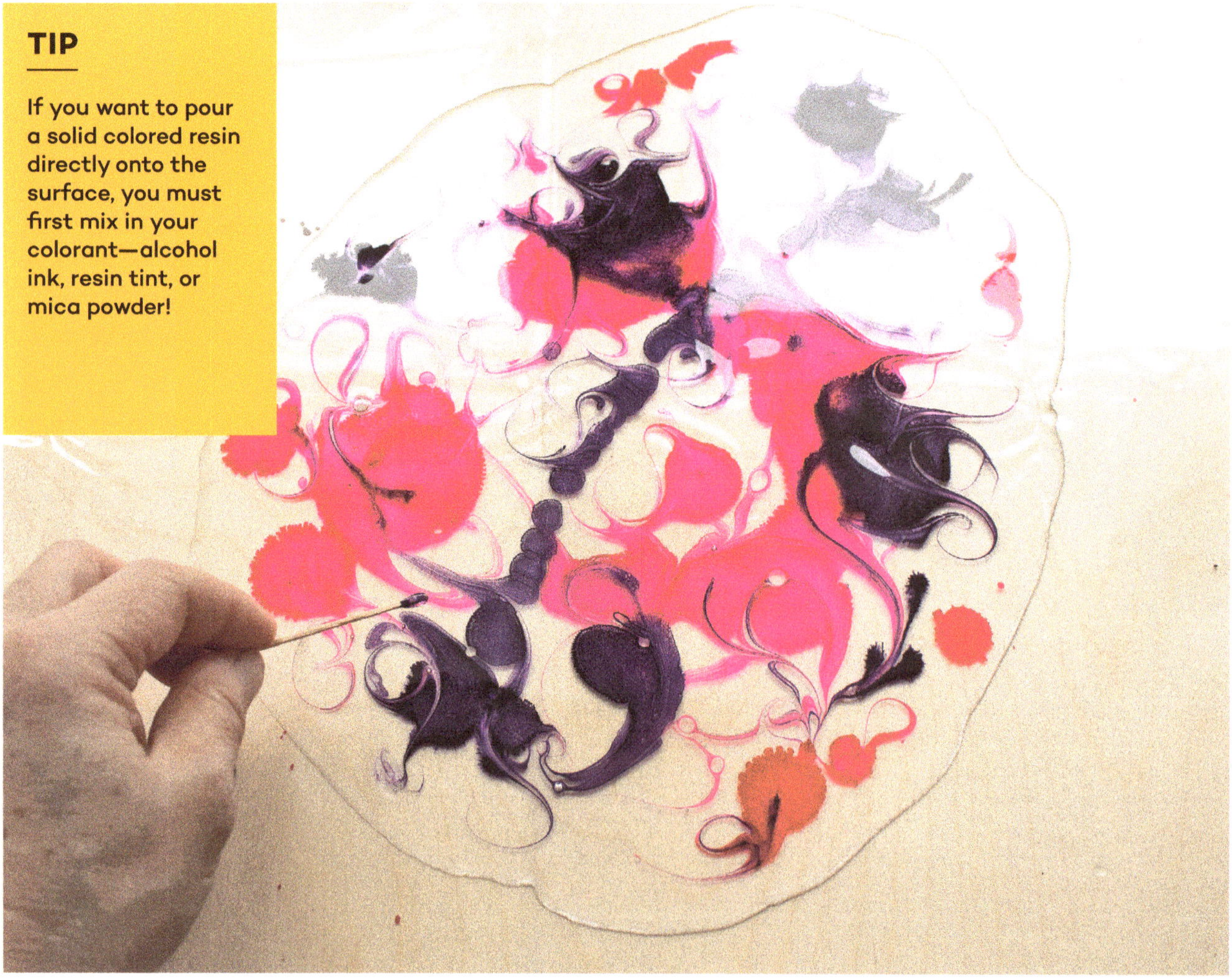

TIP

If you want to pour a solid colored resin directly onto the surface, you must first mix in your colorant—alcohol ink, resin tint, or mica powder!

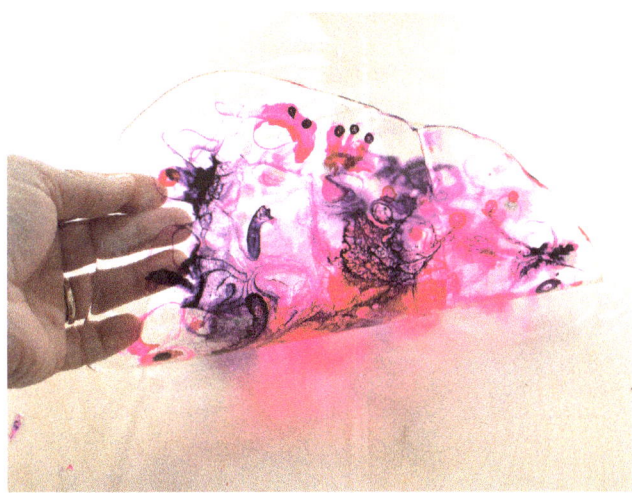

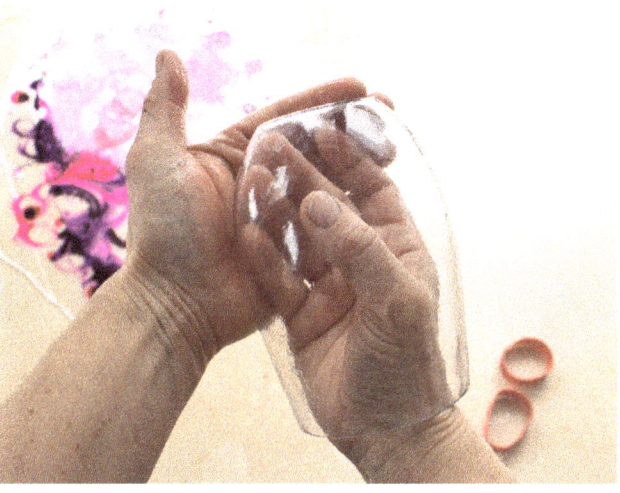

5. Allow your resin to sit covered for approximately 12 hours (this will vary depending on the brand of resin you use) or until the resin has cured enough to handle and still be flexible to manipulate. Peel the resin off of the plastic material.

6. Lightly coat the large plastic vase or container you will be using to shape the vessel with silicone oil. This will prevent the resin from sticking to your container and will allow you to remove it easily when the resin shape has hardened.

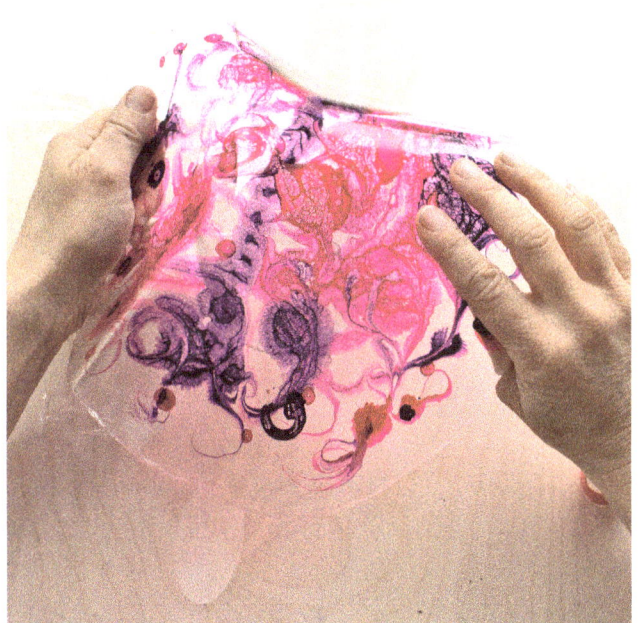

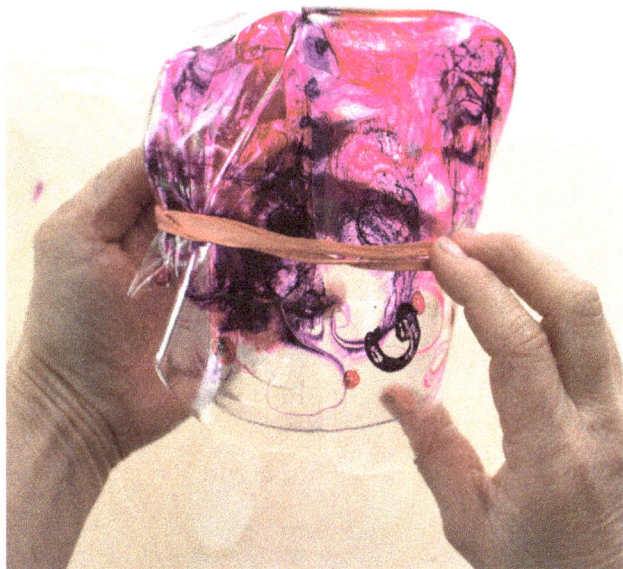

7. Place your resin over the oiled container. Shape it around the container and use some rubber bands to hold the resin in place.

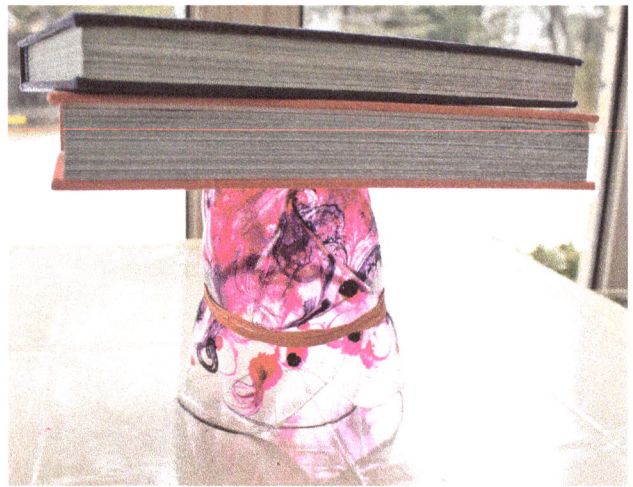

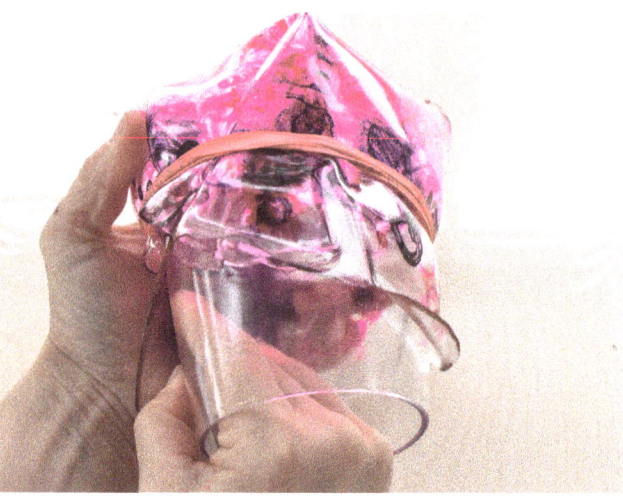

8. Flip upside down and weight the bottom of the container with some books. This will prevent the resin from lifting and sliding off of the oiled container. The books will also flatten the bottom of the vase so it will sit flush on a surface once fully cured.

9. Leave for another 24 hours before removing the books and prying the resin from the container.

TIP

Using a plastic container as your mold shape instead of a solid glass container is a better option. If your resin becomes difficult to remove, the plastic is more flexible.

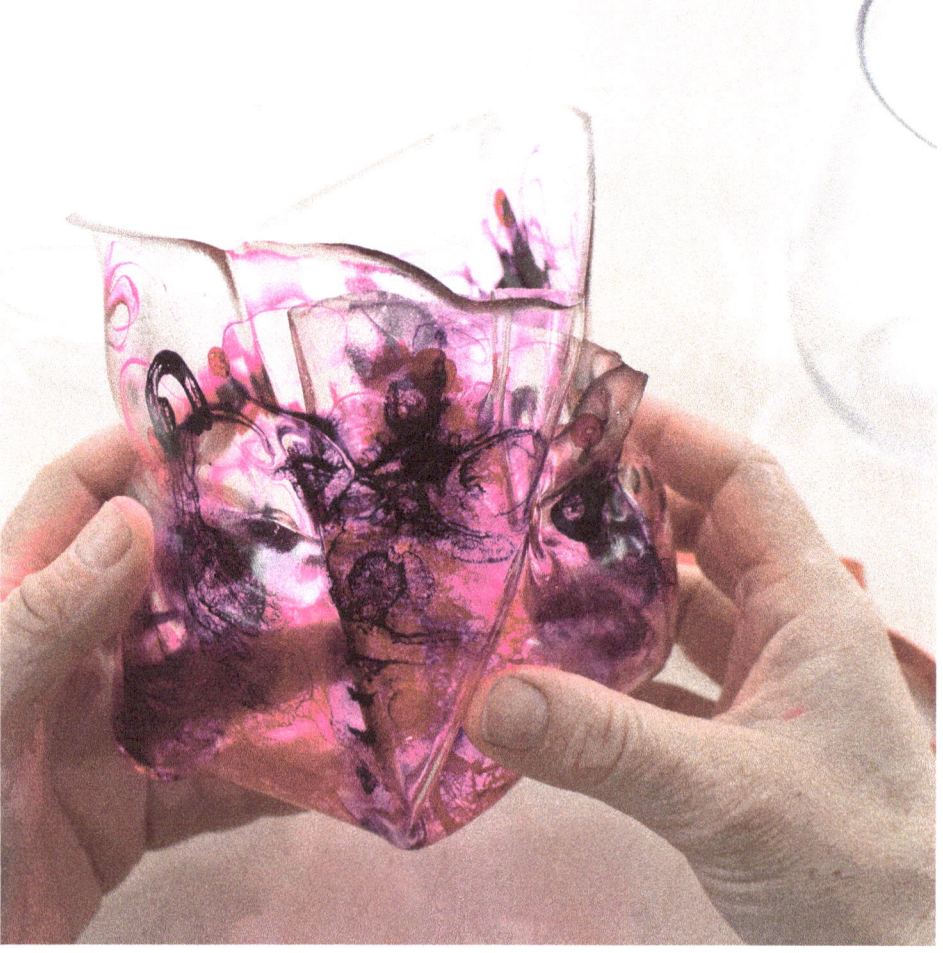

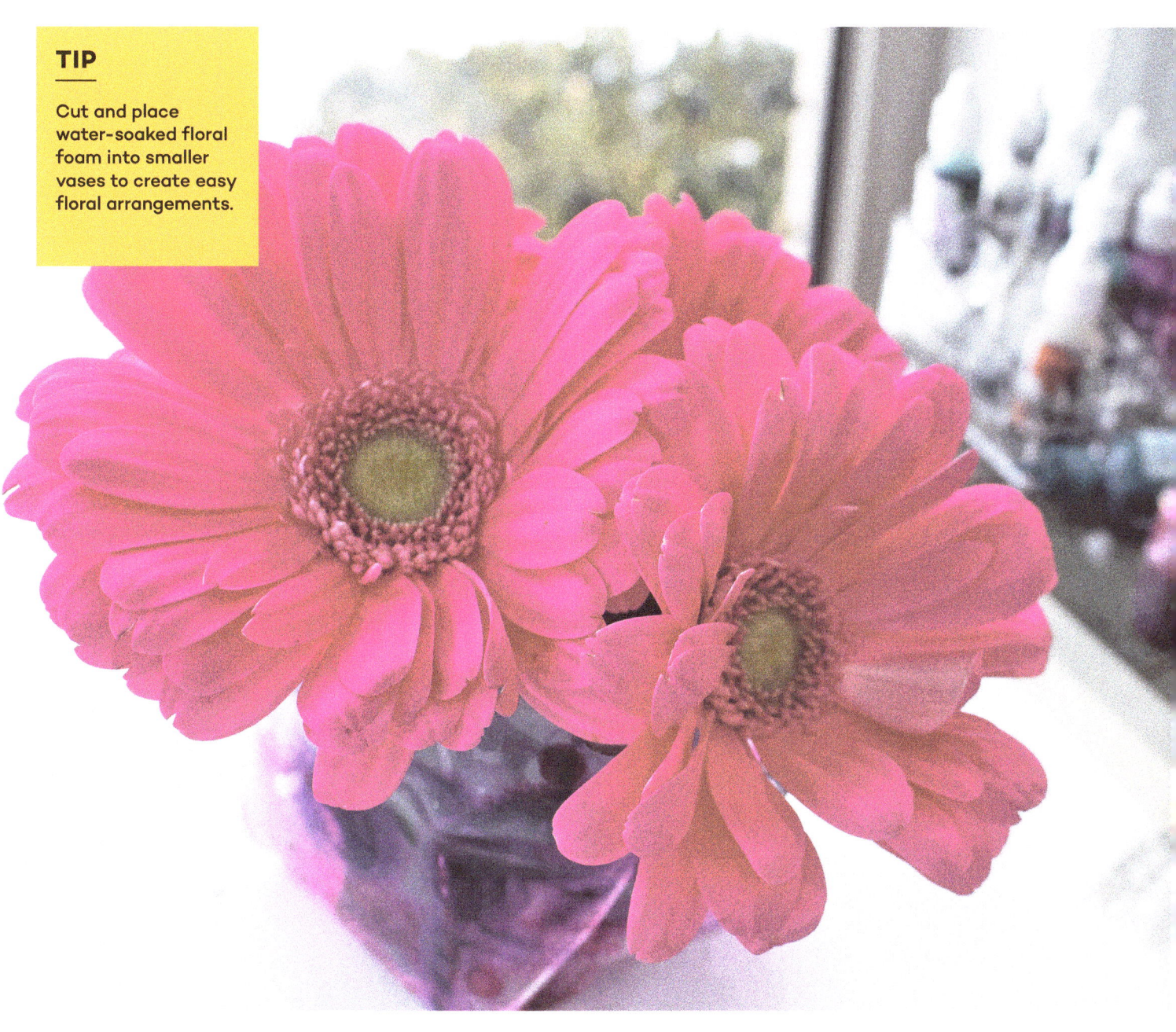

TIP

Cut and place water-soaked floral foam into smaller vases to create easy floral arrangements.

Pretty and functional. Use your vase for potted plants or fresh flowers!

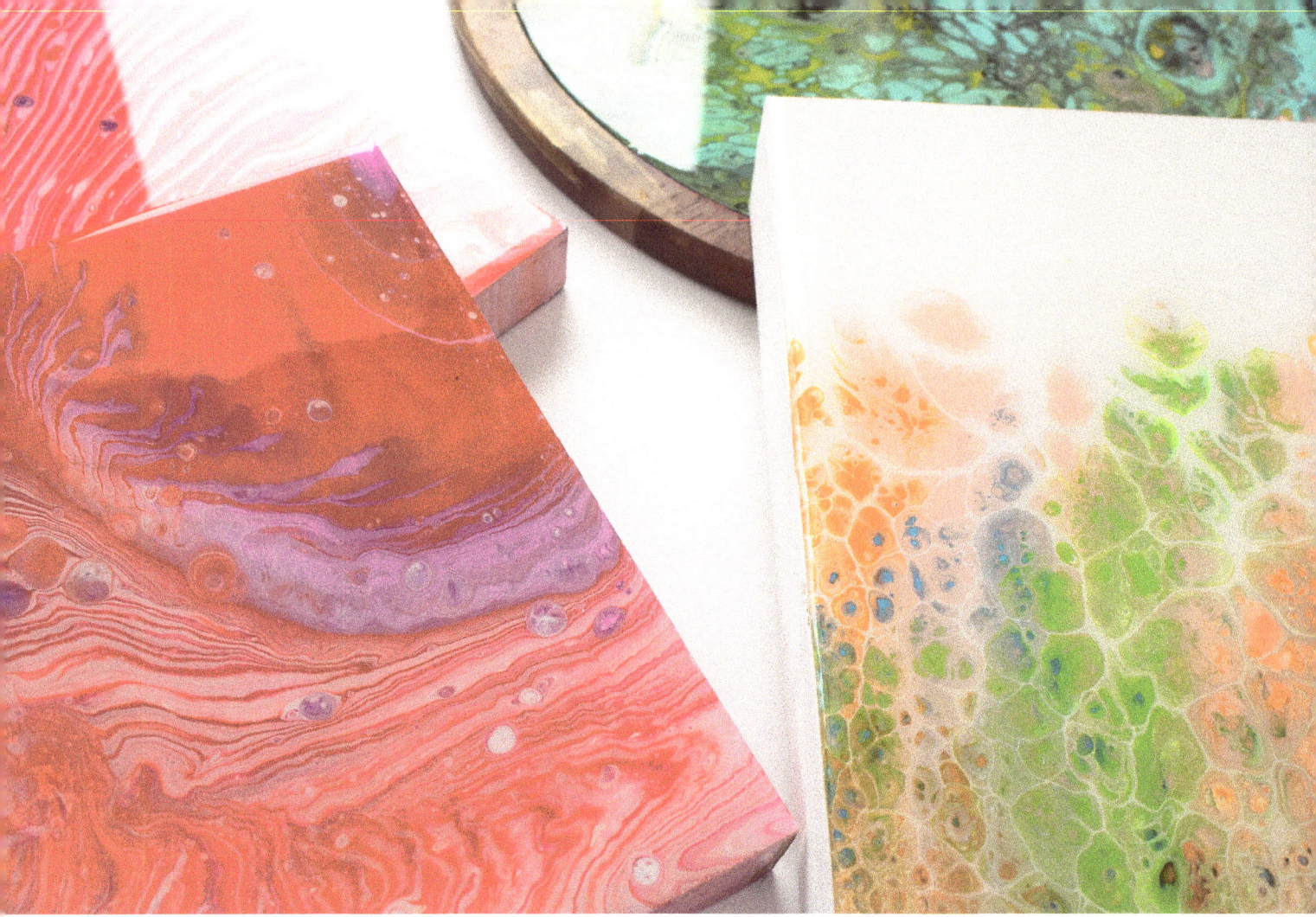

PROJECT 10

ACRYLIC POURS IN THREE STYLES

Acrylic pouring is everywhere. It is a huge trend that is here to stay. You only have to search the term on YouTube and you'll be introduced to thousands of videos on various techniques and paint formulas. The most common type of pours are swipes, dirty cup pours, and ring pours. While each of these three techniques are shown in this book, you should know that there are no set rules when it comes to preparation of a paint recipe. Everyone has an opinion on how much of one ingredient should be added to a formula. A lot depends on the brand of paint you use, the color, and the initial viscosity of the paint—soft, medium, hard, or fluid body. That said, paint pouring is all trial and error. It takes time to find a process that works for you. It is best to start off on a small scale until your technique and formula are consistent. Making mistakes in paint pouring can become costly, so take baby steps!

MATERIALS

substrate for paint

acrylic paint, including white

plastic cups for pouring

paint pouring medium*

craft sticks/spoons for stirring

silicone oil or dimethicone

piece of paper or cardstock

artist varnish

paintbrush

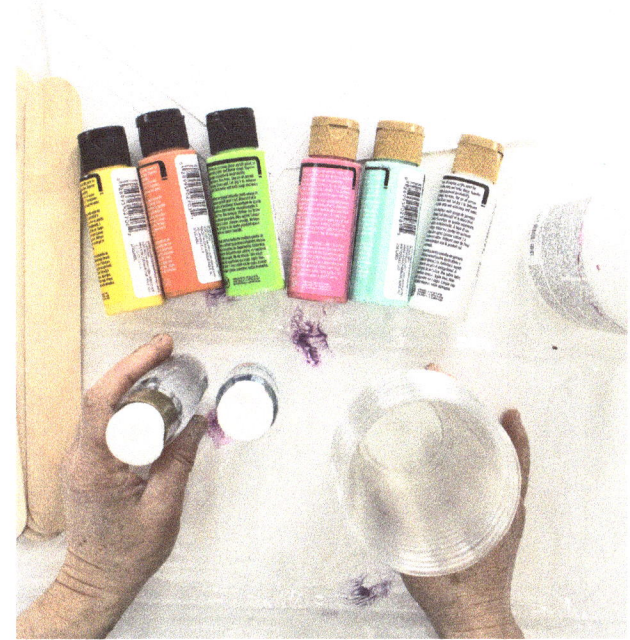

TIP

Paint pouring is very messy. Make sure your surface is well protected. Use a plastic tablecloth or set your canvas on a rack in a tray to catch the paint.

*Paint pouring medium examples: latex extender (Flood Floetrol), PVA glue and water, Golden GAC 800, Liquitex Pouring medium

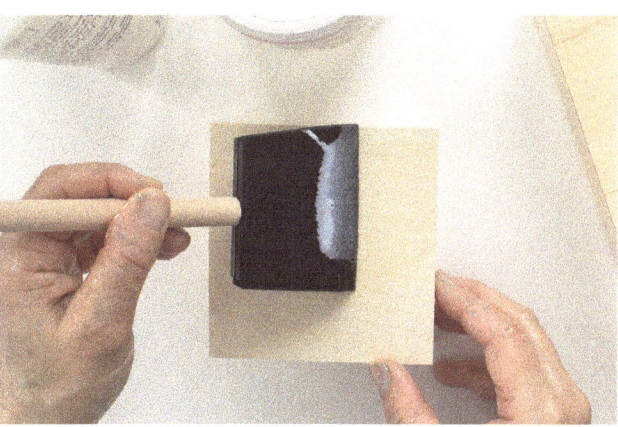

1. Think about your end result when choosing a substrate. If you plan to add resin as a top coat, then a rigid substrate such as artist wood panel is recommended. A canvas can be used, but additional backing support may be required. A pre-primed gesso hardboard is ideal for acrylic pours because the preparation is already done for you.

2. If pre-primed hardboard is not available to you or is too expensive, the next best option is to prepare a wood panel. Use a wood sealer followed by gesso and allow to dry before starting a pour. You may also want to take an extra step by brushing on a layer of gloss gel medium to reduce SID (see below).

TECH SUPPORT FROM GOLDEN PAINTS

*Support induced discoloration, or SID, describes a phenomenon that can occur when the acrylic appears to change in color upon drying. It usually takes on a yellow, orange, or brown tint due to impurities in the substrate being drawn up into the acrylic film. The discoloration occurs while the paint or medium is drying and curing and should not continue or happen after the film is cured. These impurities can be found in supports including woods, hardboards, particle boards, and some canvas and linen supports. These impurities can include glues, resins, sizing, and any soluble materials in the substrates. SID is only applicable when painting with acrylic paints and mediums and is most noticeable in thicker applications of clear or translucent mediums and gels and in opaque and semi-opaque pastes. It can affect some paint colors as well. (Source article from Golden Paints, www.justpaint.org.)

PAINT SWIPE

The key to any paint pouring technique is creating the perfect consistency. The paint cup goal is to mix up a smooth, creamy, not-too-thin, not-too-thick recipe.

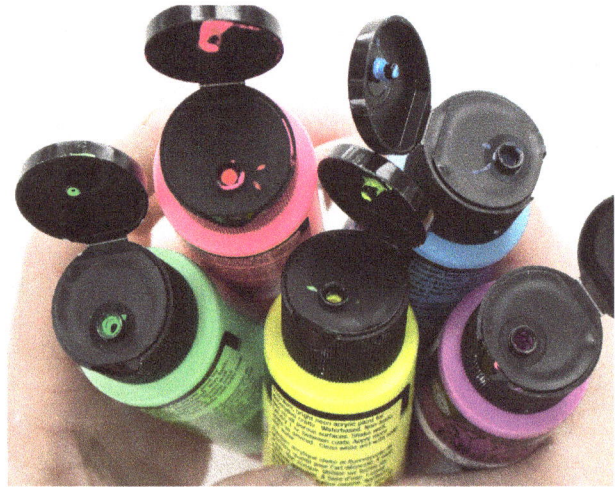

1. Begin by selecting your paint colors and adding a generous squeeze of each color to a plastic cup. You will want to add a pouring medium to your paint too. Pour approximately 2 parts pouring medium to 1 part paint. If you cannot find an artist pouring medium, you can substitute with PVA craft glue and water, found in most art and office supply stores. Stir your paint until it is the consistency of thick cream. If after stirring you find your paint still on the thicker side, add about ½ teaspoon (2.5 ml) of water and stir again. You may need to add more depending on the paint's viscosity. If you started with a heavy body paint (which is much thicker than a soft body), then you will need to add more water than you would if using a fluid body paint. Make sure you read the labels on your acrylic paint so you know which type you've purchased. Different brands of paint will also yield different results. A cheaper student-grade paint will not act the same as a high-end professional-grade one. It's all in experimenting! Hence my comment about there not being a simple one answer paint formula.

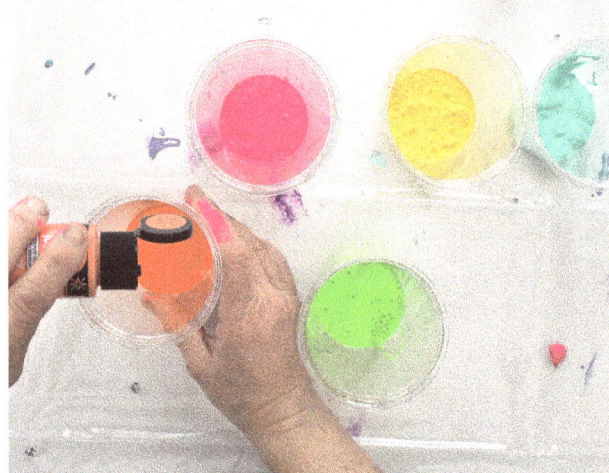

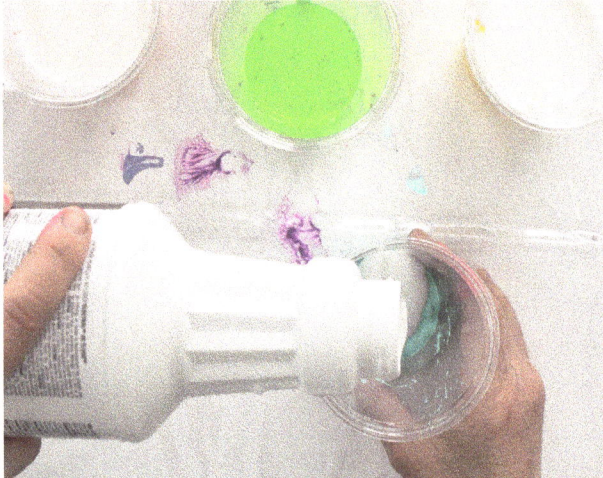

TIPS

- Use a fluid body medium that doesn't need to be thinned out as much with water and latex extender.

- You can use PVA glue and water as a pouring medium substitute. Mix 2 parts glue to 1 part warm water. Mix thoroughly until smooth and then add the amount needed to your paint.

- Swipes and dirty cup pours are created with cells. This is the main reason people love doing this form of art.

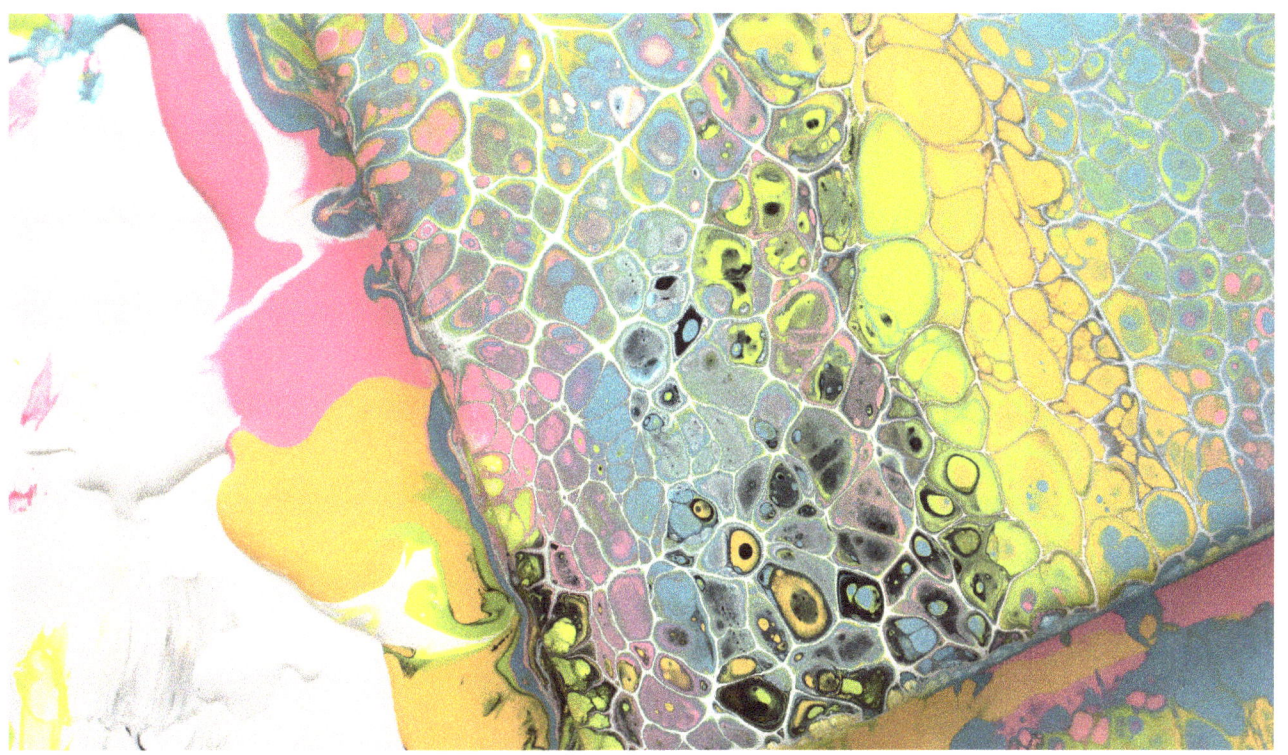

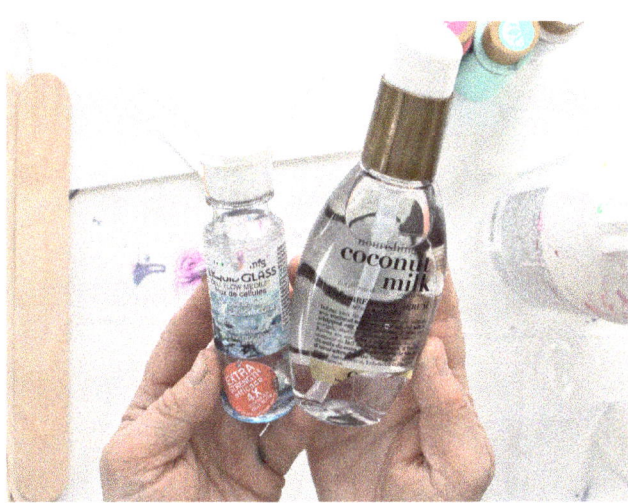

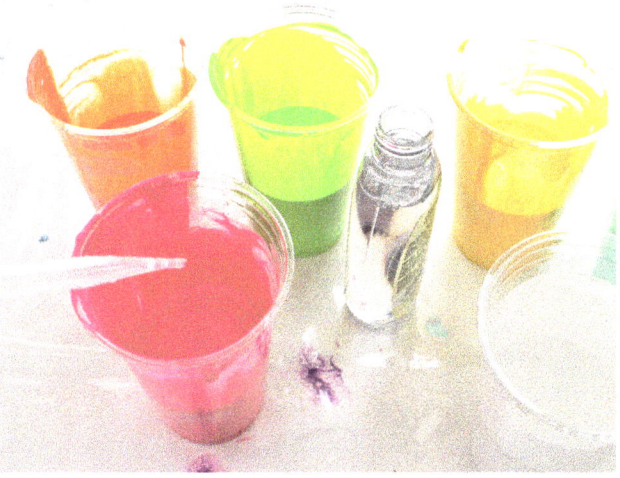

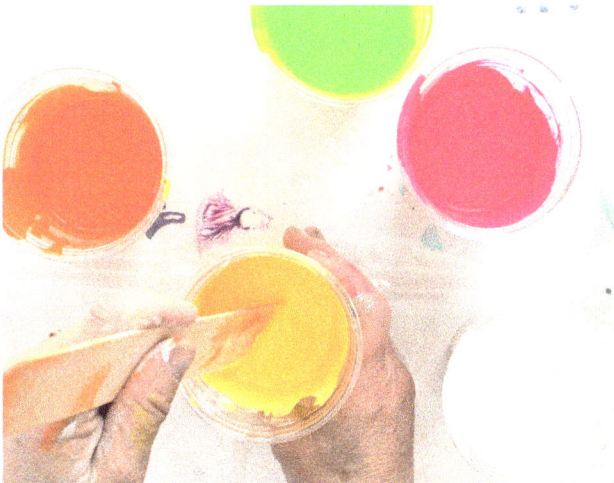

2. While the paint on its own will provide some small self-made cells, the addition of an artist silicone oil or product containing dimethicone (hair serum) will produce a lot more. Add 1 to 3 drops of silicone to each paint color *except* your swipe color, which will be white. Lightly stir your paints.

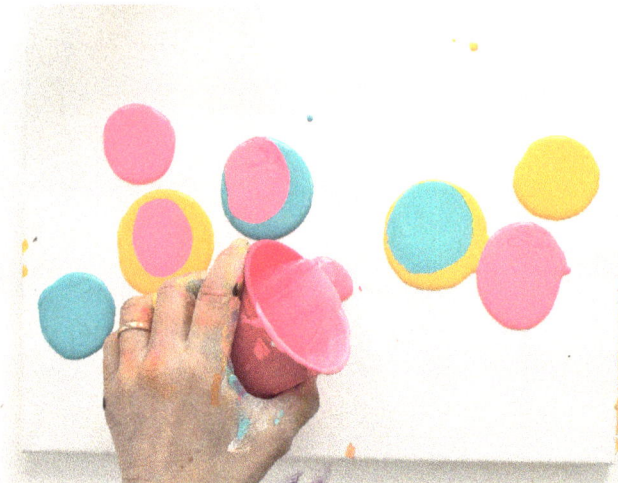

3. If you plan on just varnishing your art for protection over resin, a pre-primed canvas will be fine. Now the fun begins! Start by pouring little circles of paint randomly across the canvas, alternating colors and dropping some into other circles. This will produce multicolored rings around the cells themselves.

TIP

Make sure your substrate is level to keep the paint from running in one direction.

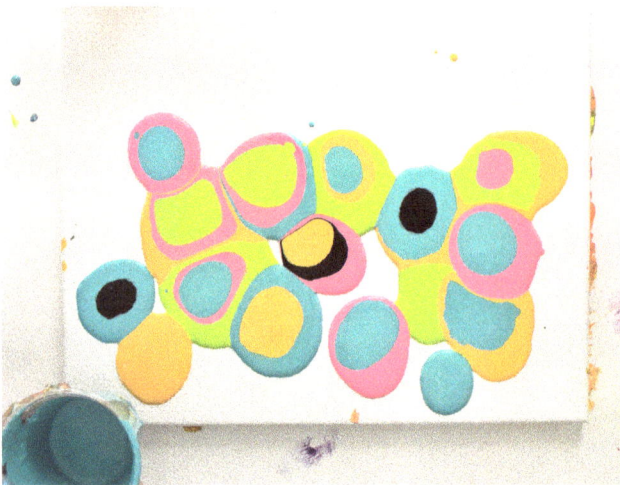

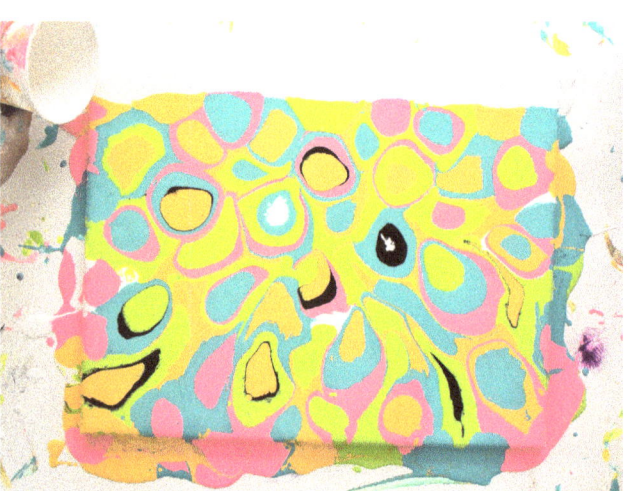

4. Continue across the canvas until you have filled in most of the surface area. You can pick up the canvas and slightly tilt it to ensure all areas on the sides are covered. Leave a little bit of space at the top where you will pour your white swipe color. Add a pour line of white paint horizontally along the canvas.

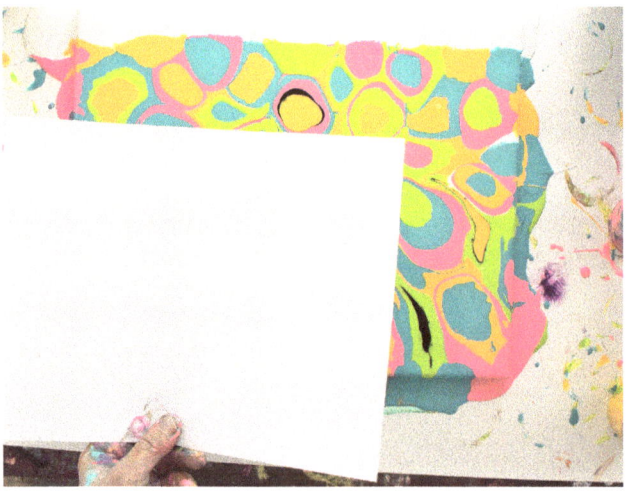

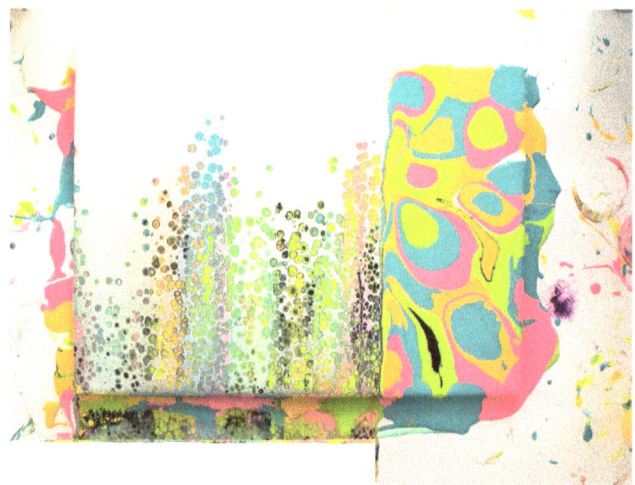

5. With a piece of paper or cardstock, gently hold and lay the paper directly on the top of the white surface, just in from the edge of the canvas. In one smooth swoop and without stopping, drag your paper across the paint to the other side.

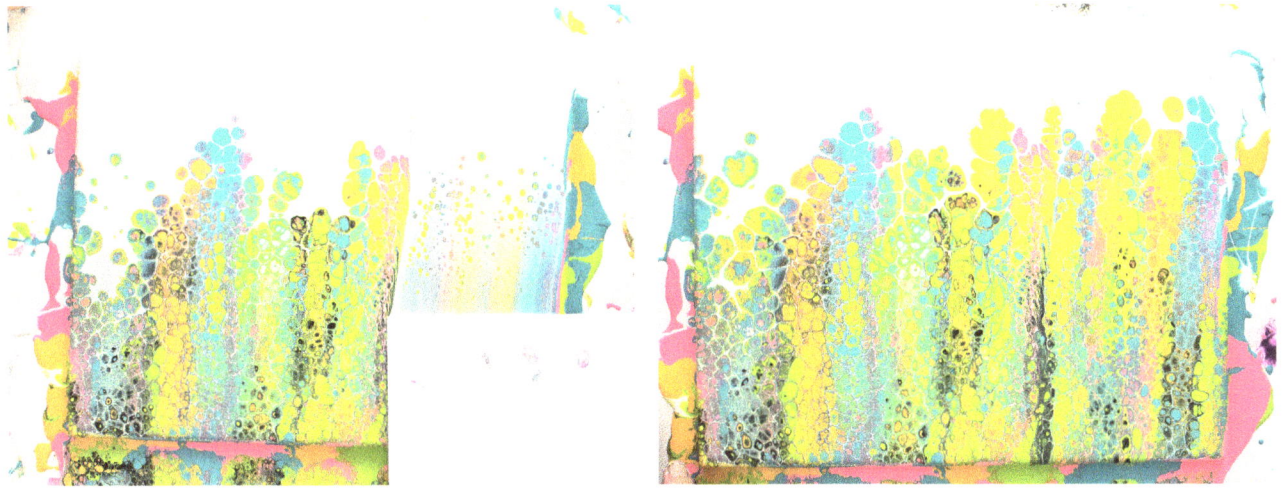

6. Wipe off the excess paint from the paper, flip it over to the clean side, and repeat the process on the second half, overlapping slightly from your first swipe.

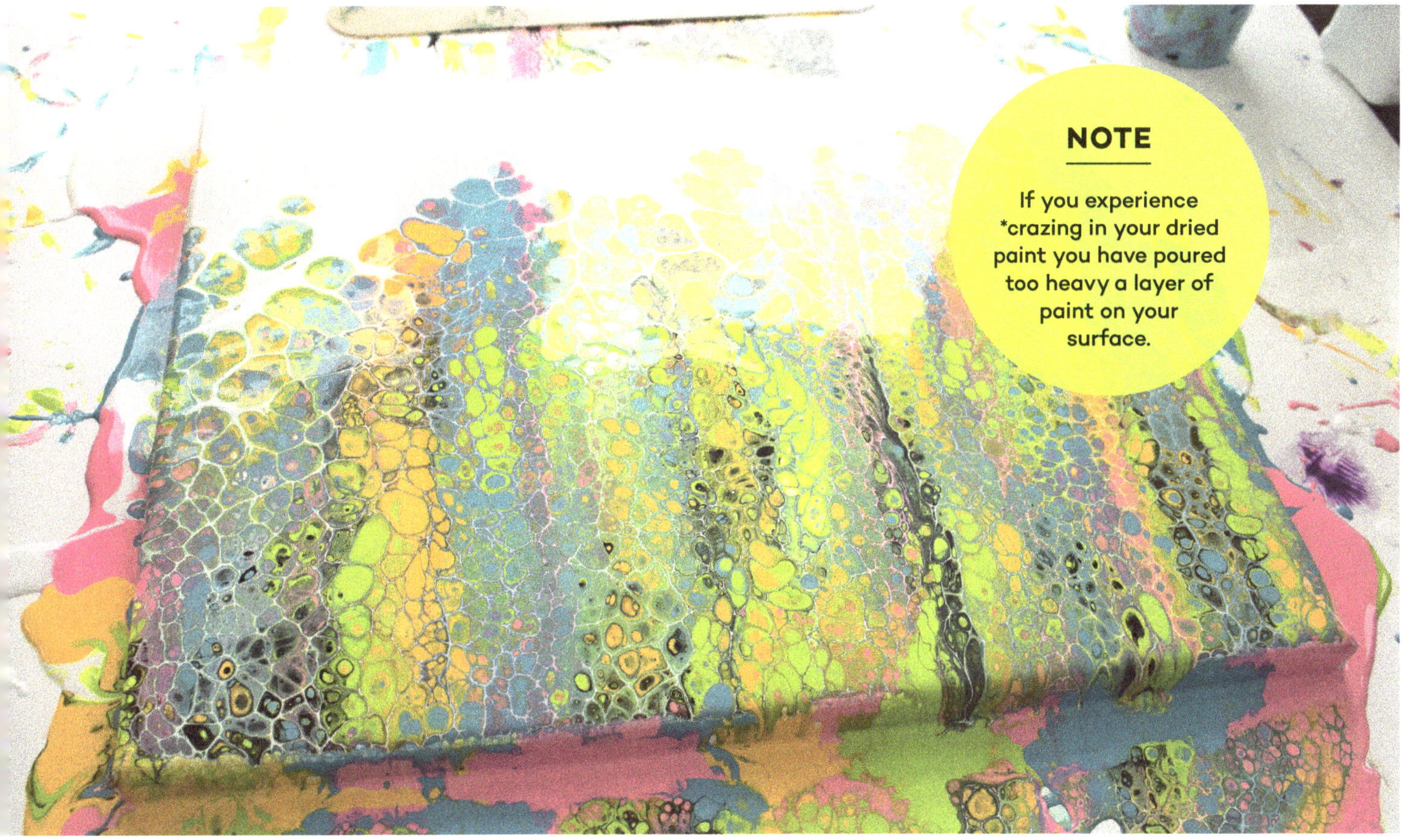

NOTE

If you experience *crazing in your dried paint you have poured too heavy a layer of paint on your surface.

7. Cells will continue to form while your canvas dries. Depending on the amount of paint you used, the consistency, and the paint to pouring medium ratio, your cells may shift or run upon drying. Outcomes may change, so experimenting and having fun is all part of the process. Once dry, protect your work with a brush-on artist varnish. (Remember to always test products first before applying them to your final artwork.)

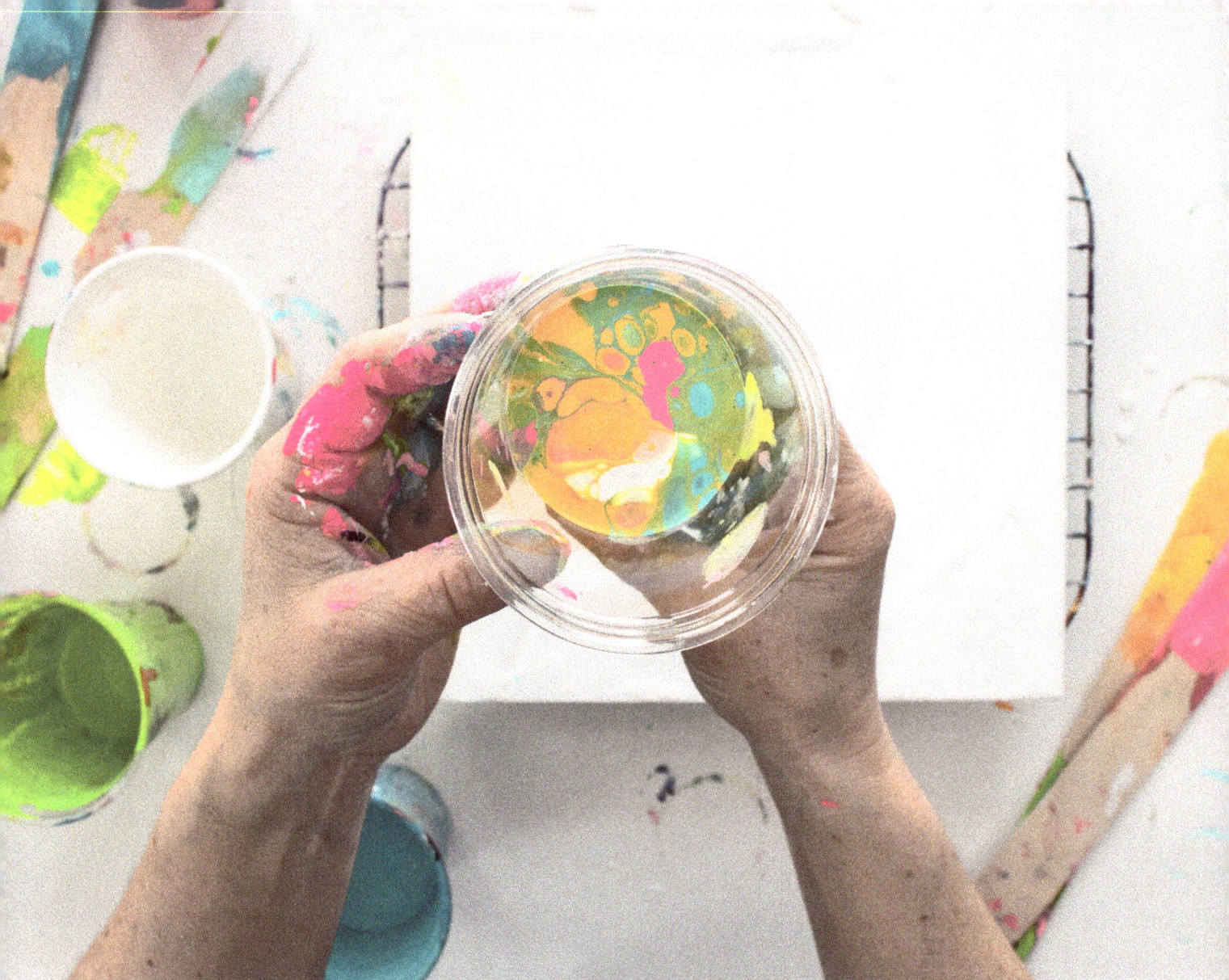

DIRTY CUP POUR

Next you will create a dirty cup pour. It is a great name for this style of fluid art, as it represents all your leftover paints that are literally poured into another cup, which is flipped over onto your canvas and lifted away, revealing a multicolored spreading of paint cells. Every flip turns out completely different, and you never know what the outcome will be. It's almost impossible to replicate a pour. So, if you make one you love, cherish it! Watching the paint transform before your eyes is addicting. It's all about what's in the cup!

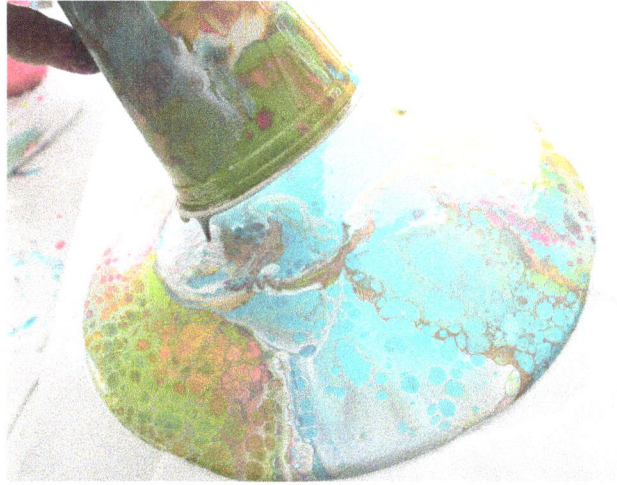

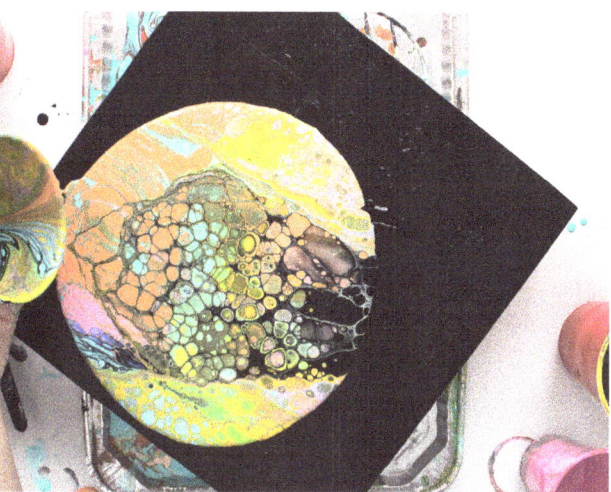

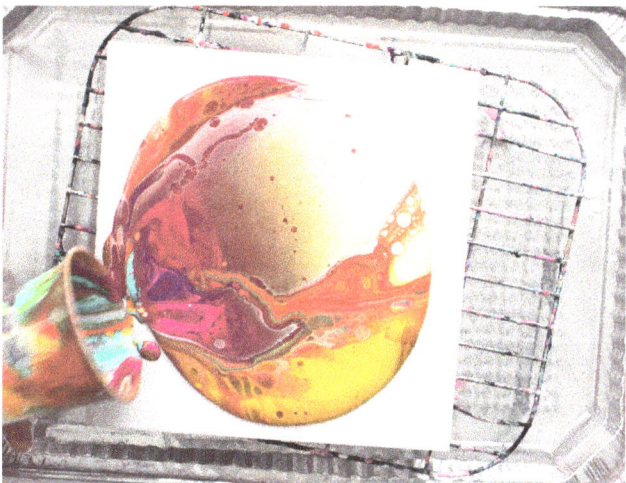

1. Begin with a clean cup in which to pour your paints. If you don't have leftover paints from your last pour or would like to use different colors, make up your new paints by following the steps from style 1.

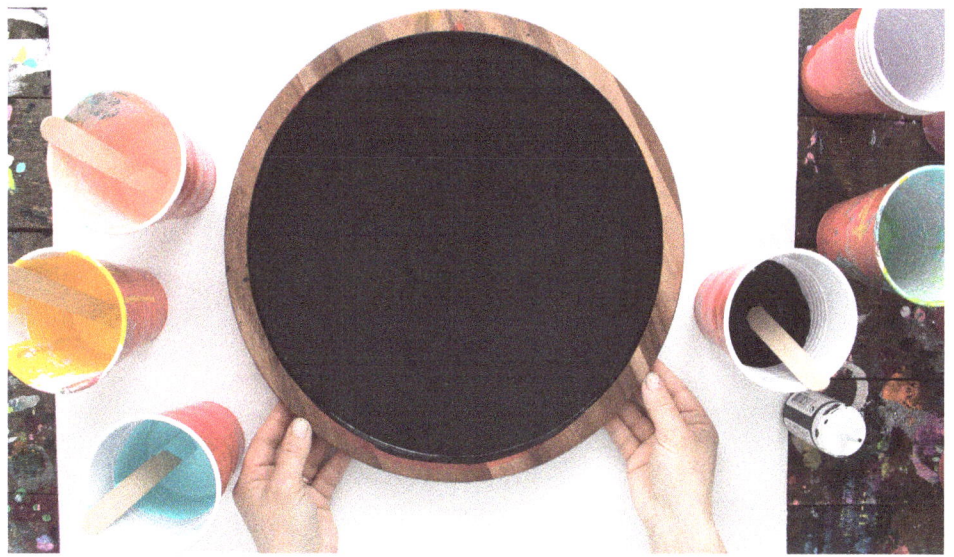

TIP

Lipped or raised-edge substrates are great because the paint doesn't run off the edge and is easier to control. Remember to work in thin layers. Too thick a layer will crack as it dries.

2. If you plan to coat your finished art in resin, use a sturdy substrate such as pre-primed wood panel, hardboard, or even a tray, which makes for a great piece of functional art!

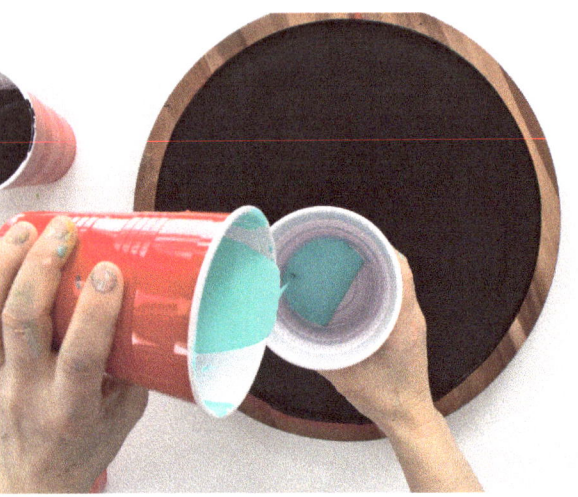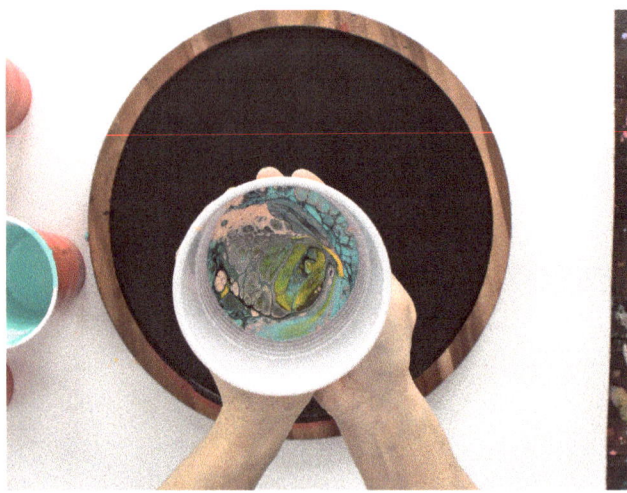

3. Begin by pouring some of your colors into a cup. Don't add too many or you'll end up with a lot of brown in your pour, especially if you use green and red, which make brown when mixed together. Alternate the paints until you have enough to cover and spread across your surface without having to tilt it too much. Before you flip it, your pour cup will look like it's starting to form cells.

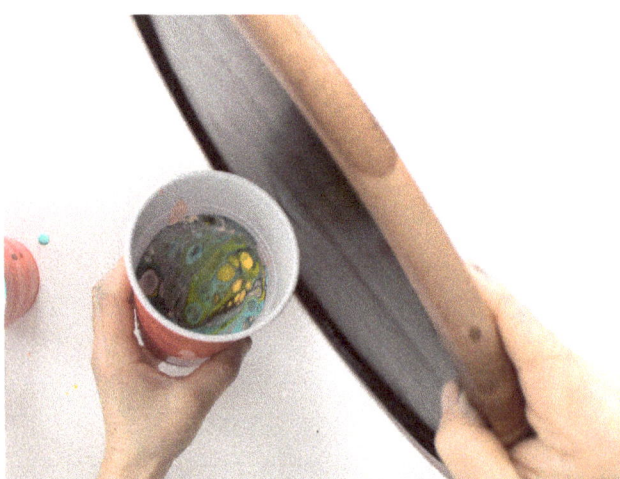

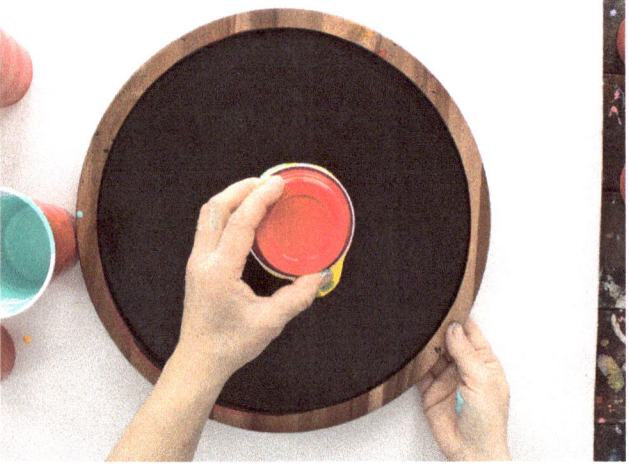

4. Hold the paint cup with one hand and use your other hand to place the tray or substrate over the open cup. Flip it over, holding it in place to rest upside down on the substrate. Wait a few moments to allow the paint to sink down the cup before pulling away the cup and releasing it.

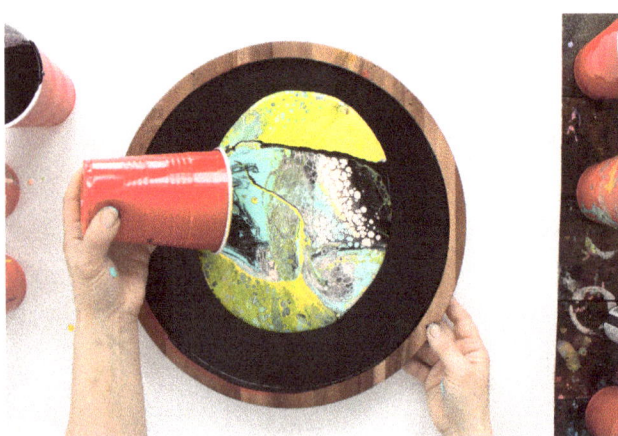

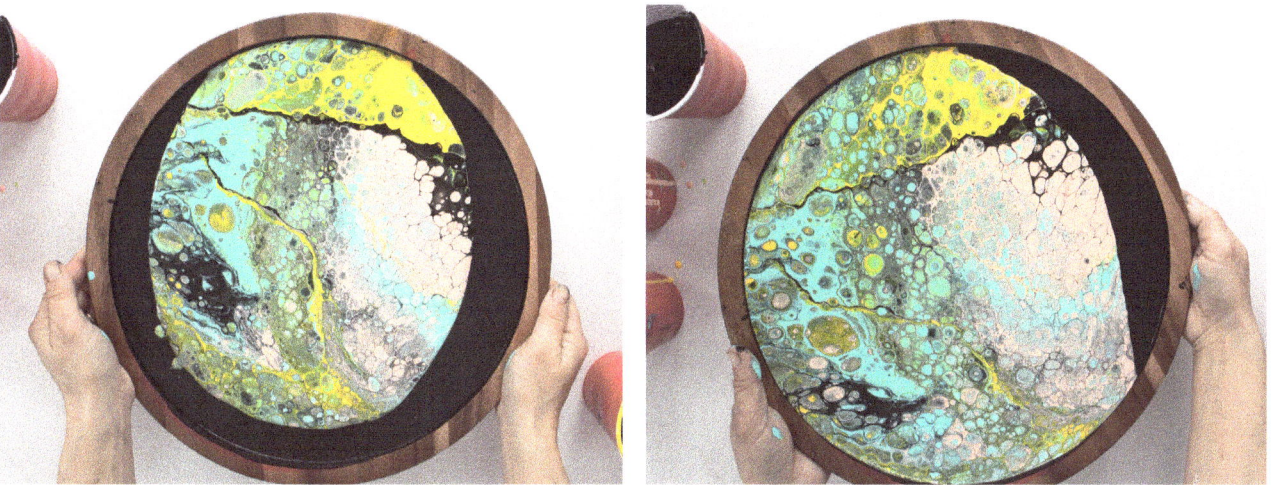

5. Pick up the substrate and begin to tilt it very slowly so the paint gradually covers the entire surface. Moving it slowly helps the pattern stay intact.

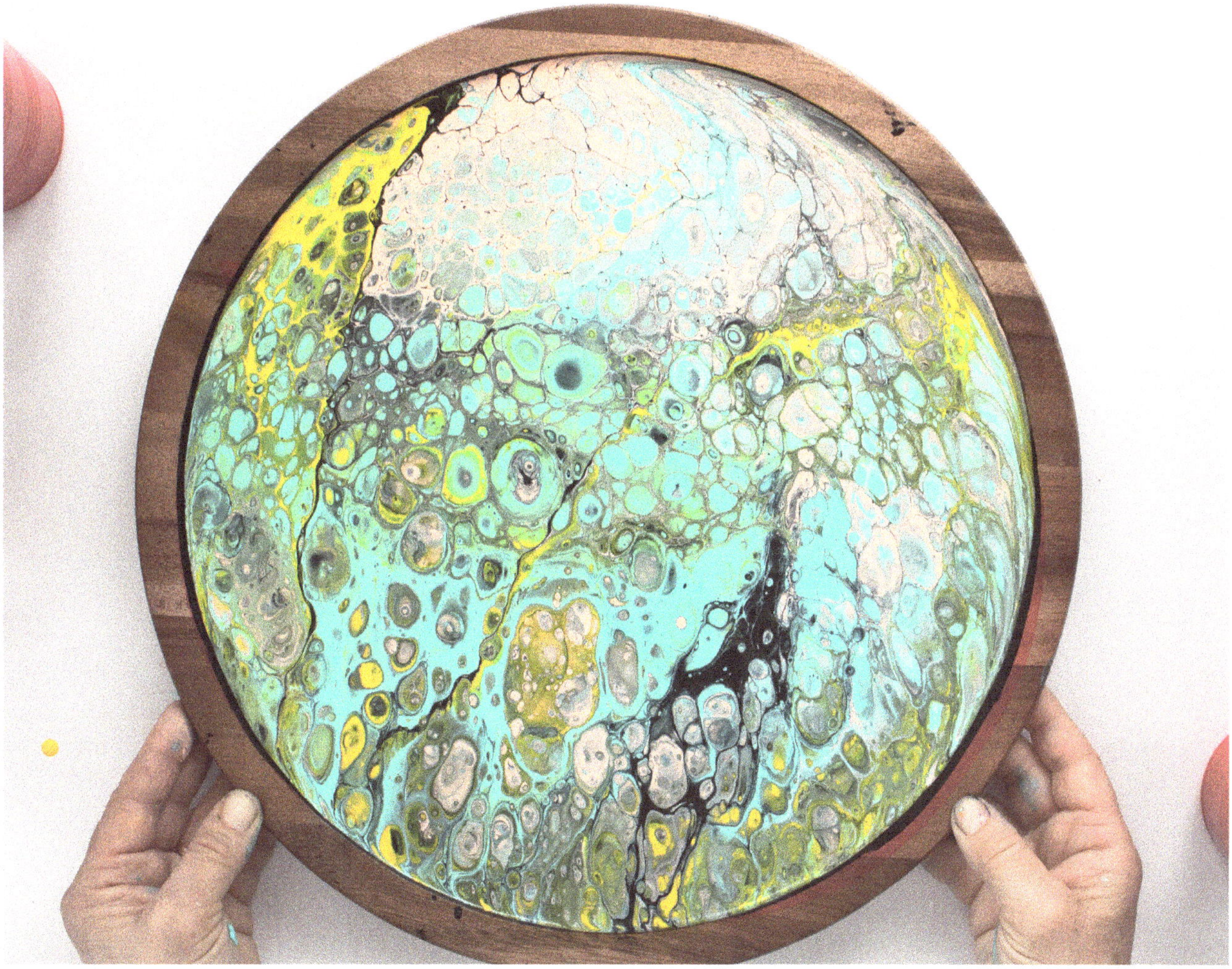

6. Allow your paint pour to dry. This may take up to several days because acrylic pours of this nature are slow to dry. Once your art has completely dried, seal and protect your work with a brush-on glossy varnish (not shown).

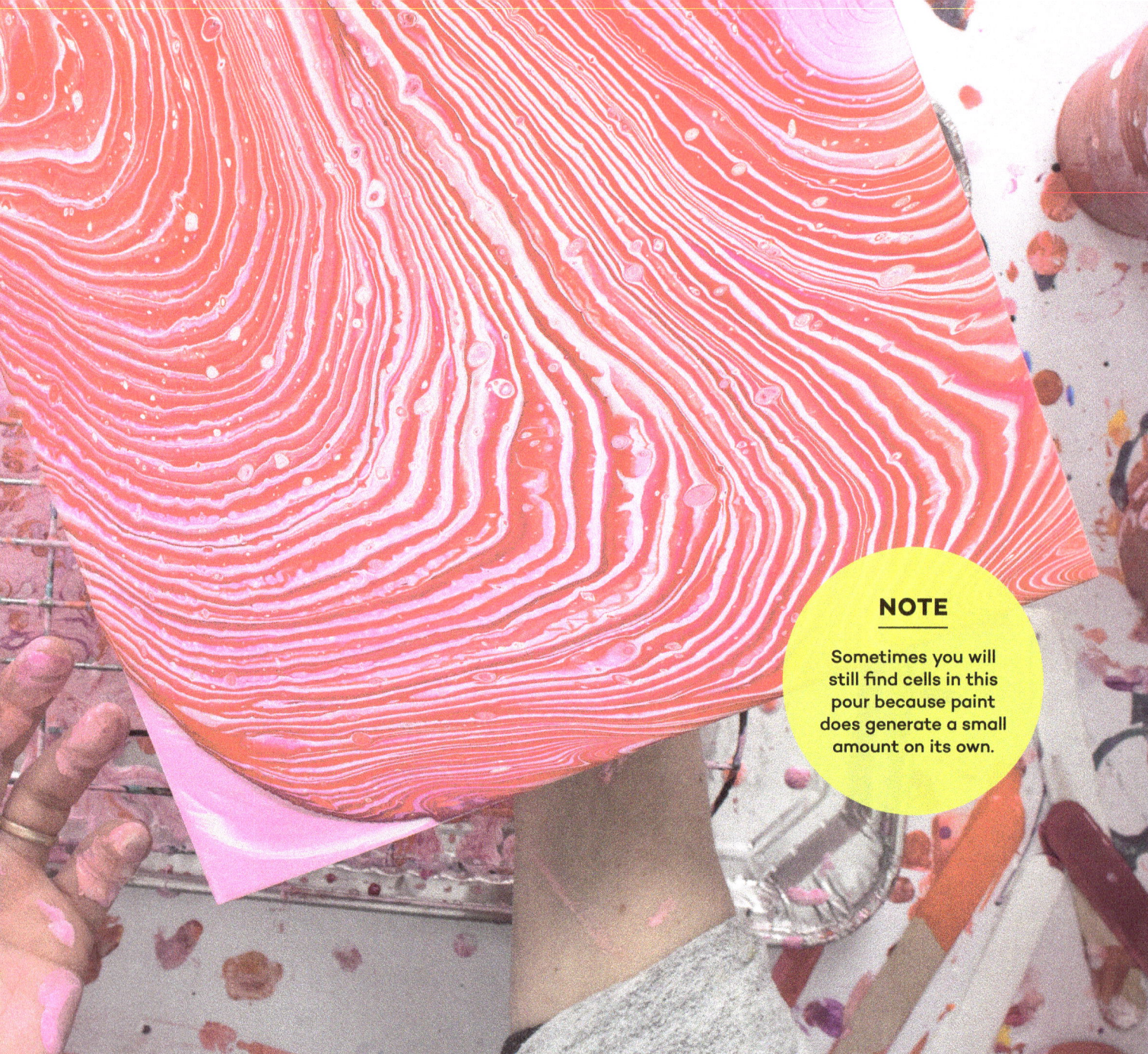

STYLE 3

RING POUR

This third style of acrylic pouring is called a ring pour because the effect is similar to the inside of a tree and its rings of life. Only a couple of colors are needed to achieve this look. The pour can be done all at once or broken up into several pours from the same cup, changing the look. You will prepare the paint in the same way as you did with the two previous styles, except you will leave out the silicone oil. You don't need to generate cells in this pour, as it's all about the circular ring shape.

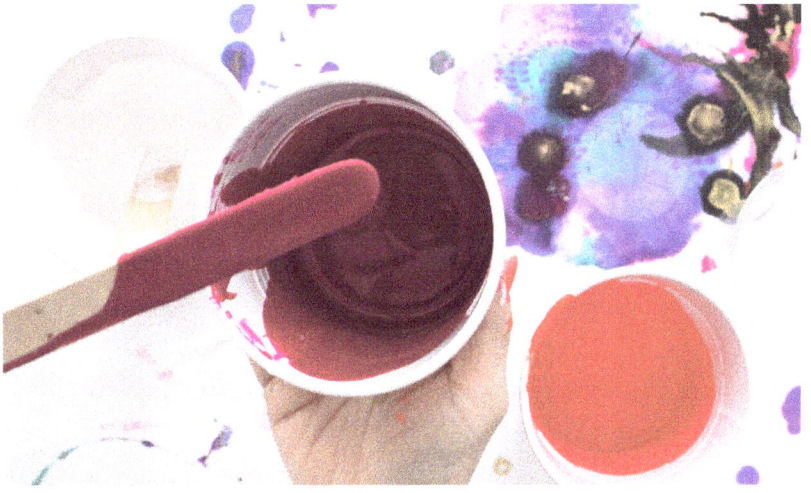

1. Select a couple of colors (contrasting ones work best with each other). If you haven't mixed your paints until this project, go back to the first acrylic pour technique and read about the steps needed to make your paint formula.

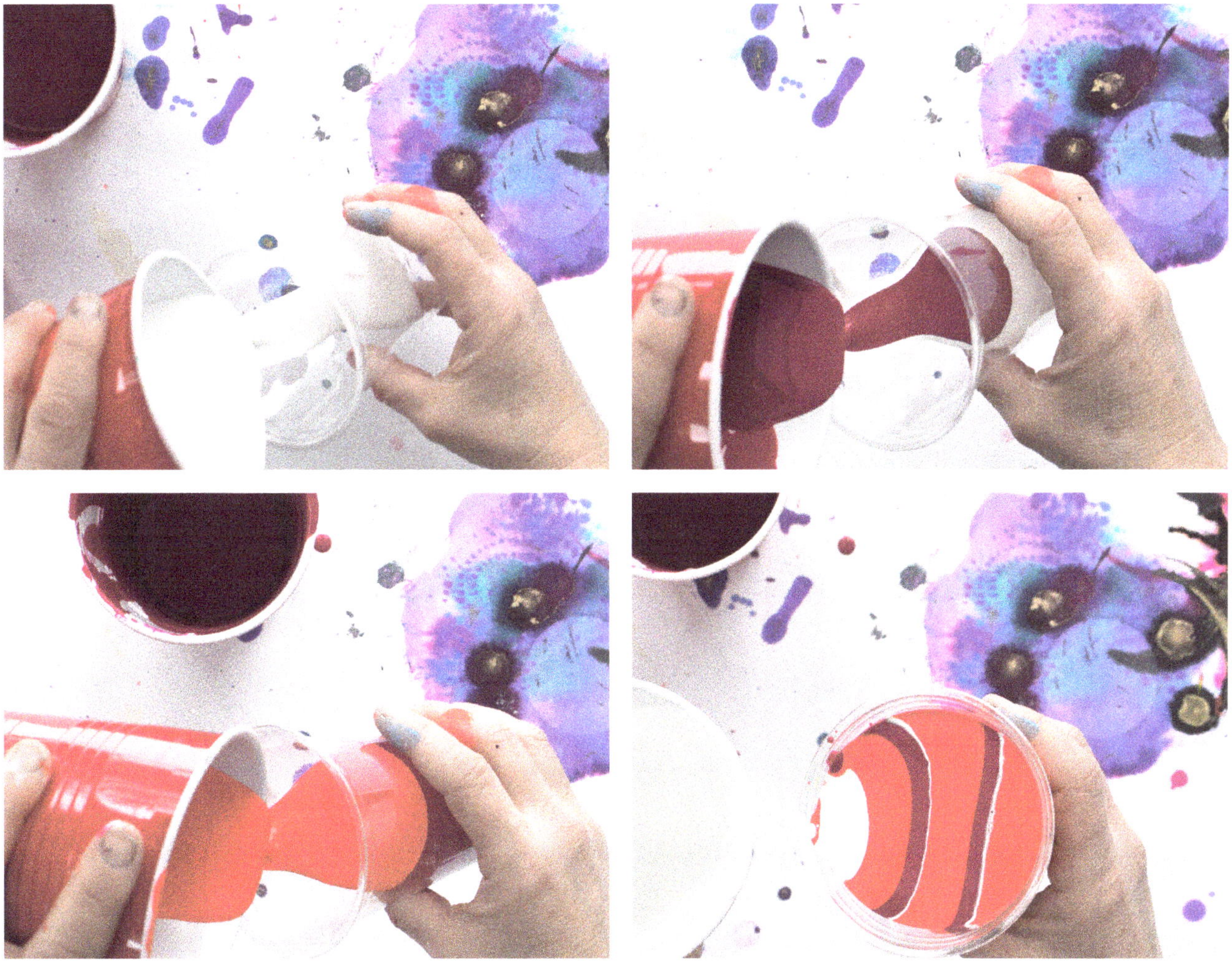

2. The key to the ring pour is in how you place the paint into the pour cup itself. In order to achieve a ring effect, pour the paint colors slowly at an angle down the side of the cup. Repeat this process with each alternating color until you have enough in your cup to begin pouring.

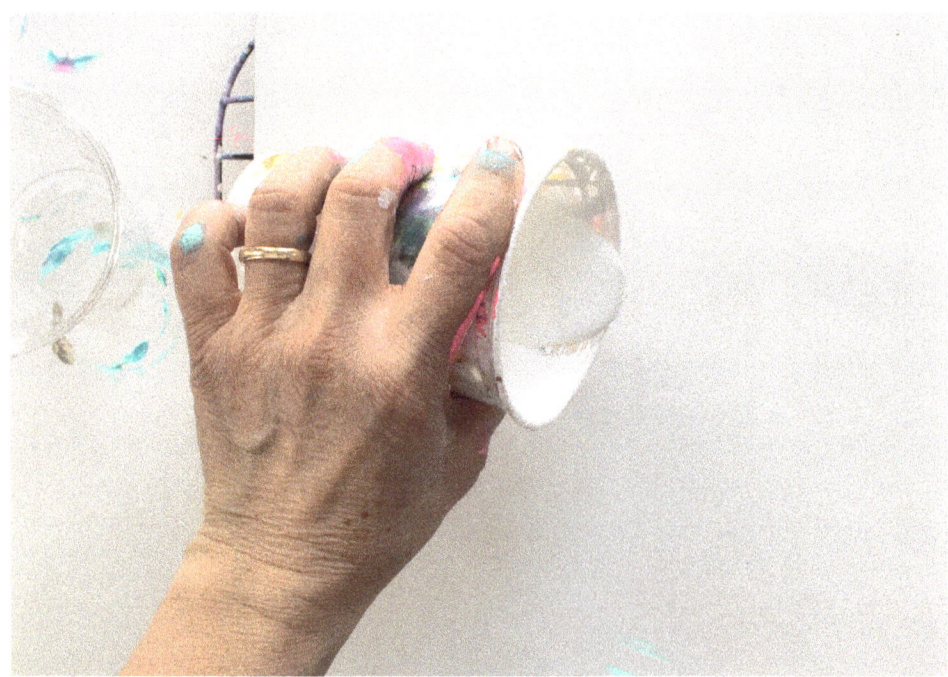

TIPS

- To help the paint spread easier across the substrate, begin with a wet layer or base paint on your surface. This creates a slick surface on which the paint can glide.

- Paint pouring is messy. Set your art board or canvas onto a wire rack with a foil pan underneath to catch runoff.

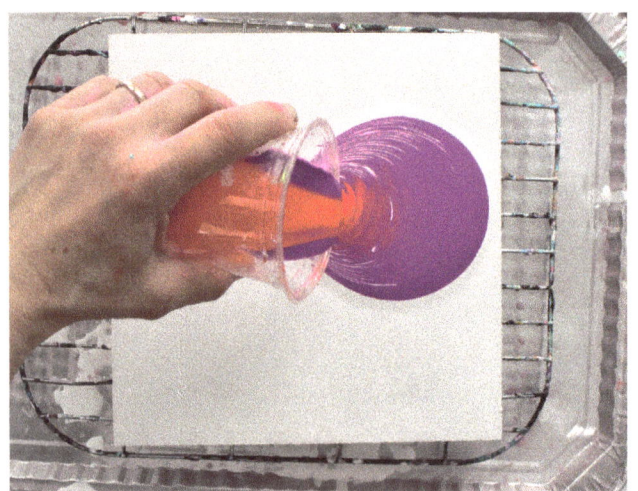

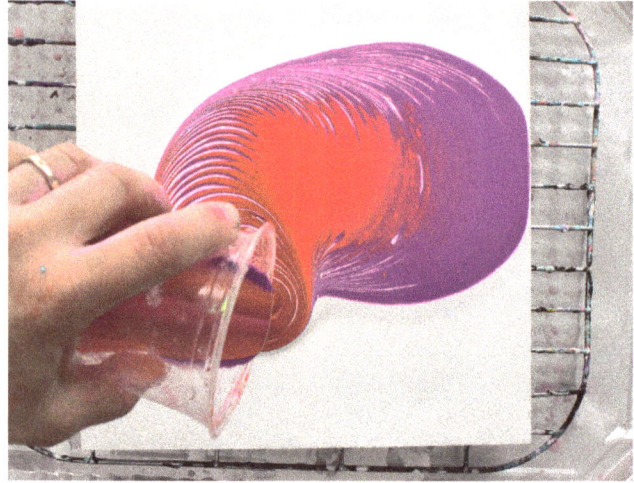

3. With your surface prepped with paint, begin pouring in a consistent small circular motion. Continue in the same area or gradually move across your surface without stopping until all the paint is gone. Or, stop halfway through and begin again on another area until the paint is gone.

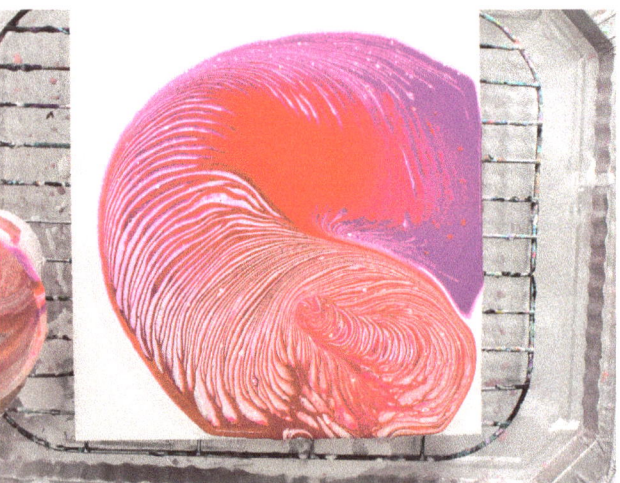

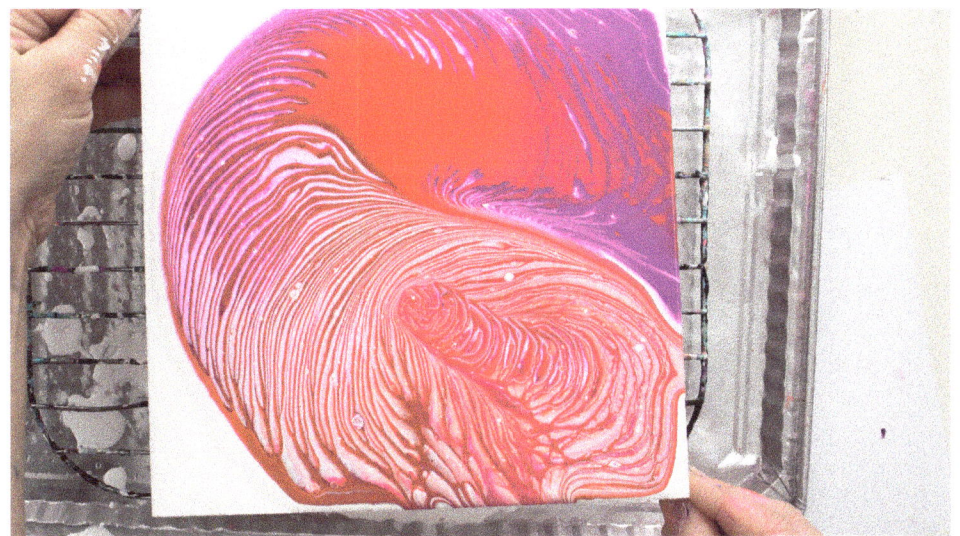

4. Carefully pick up the canvas or substrate and slowly begin tilting it to shift the paint until it covers the entire surface. Pay close attention to the edges and sides.

5. When your surface area and sides are covered, set the artwork back down on your rack, cover, and allow to dry.

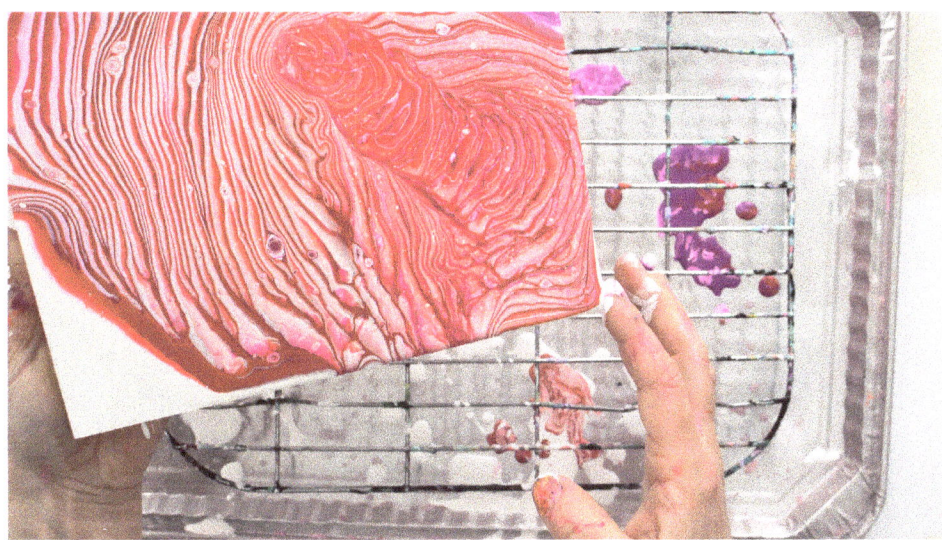

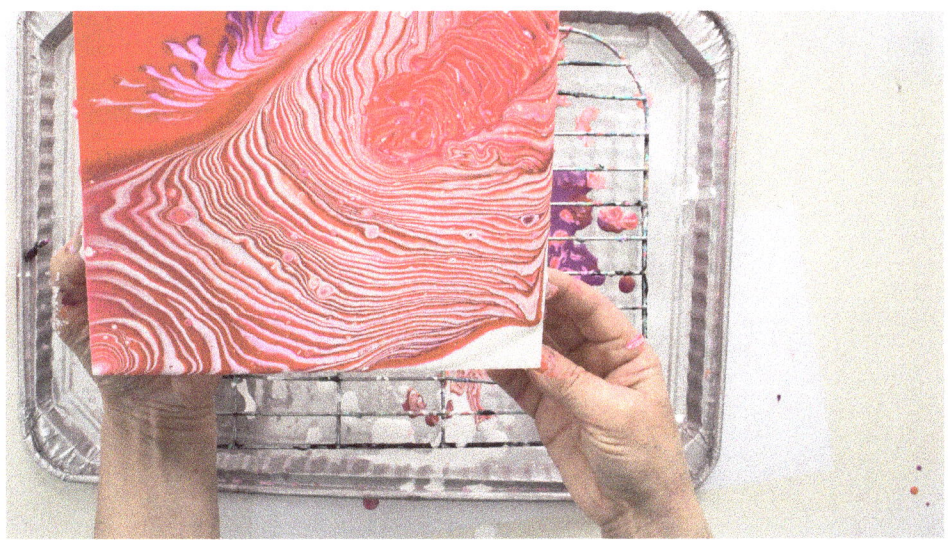

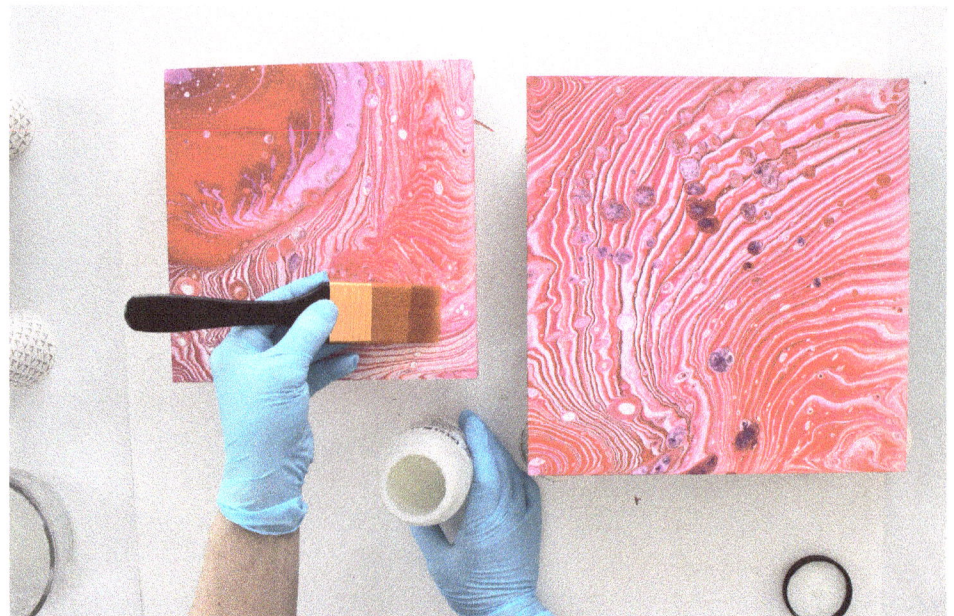

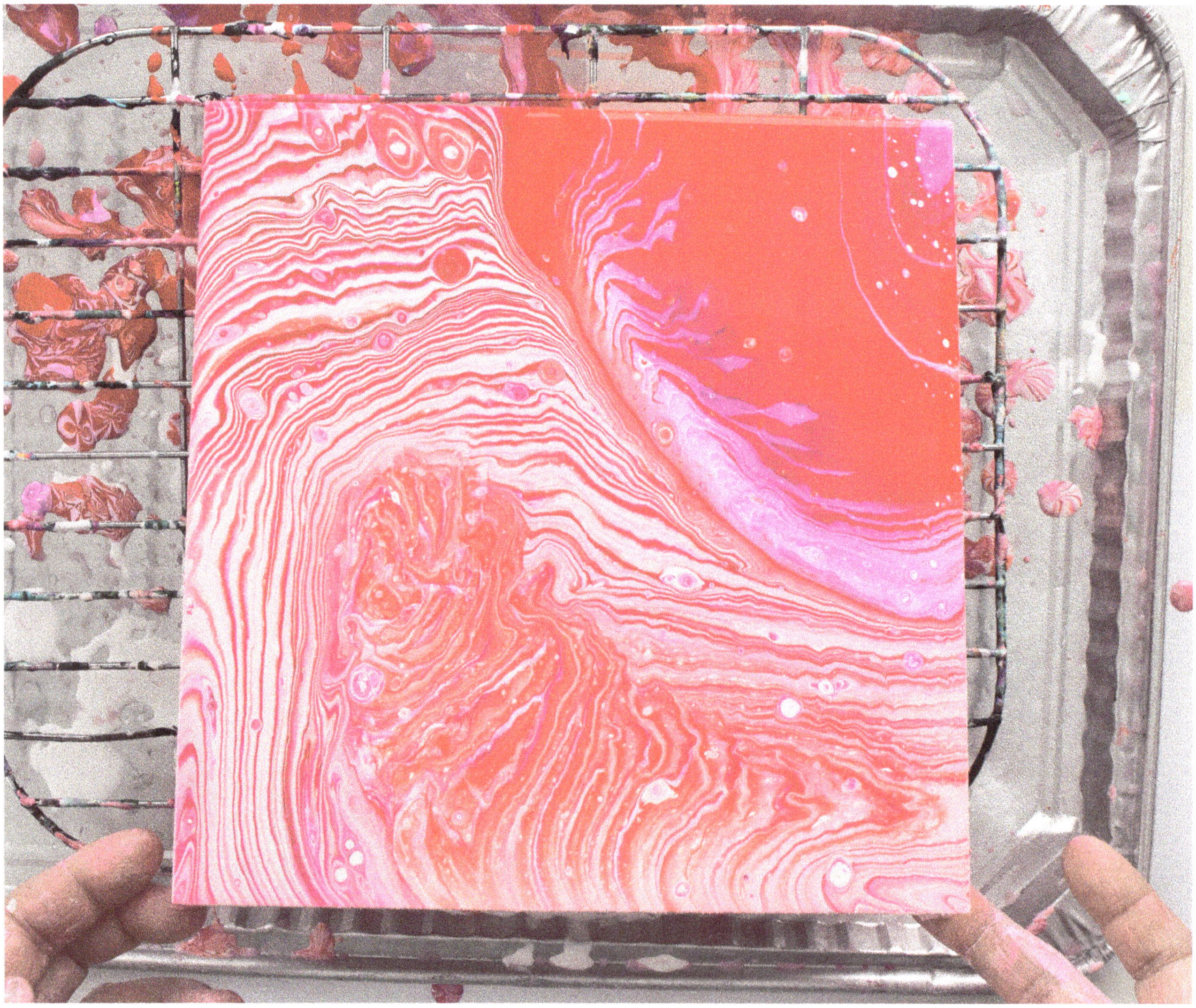

6. Finish your art pieces with a protective varnish and resin for a brilliant shine!

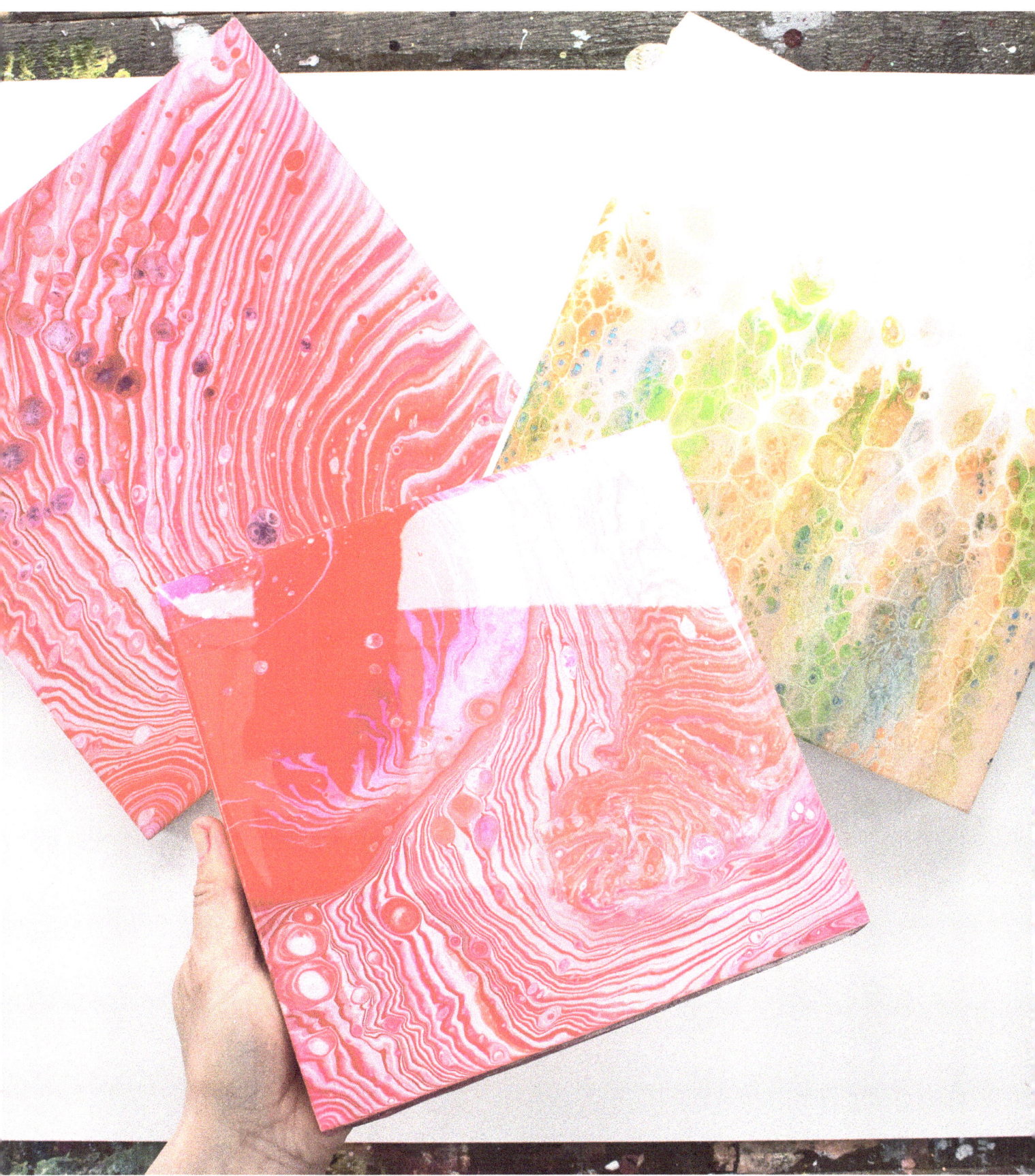

GEODE RESIN PAINTING

You've already made some agate-style coasters, now how about a geode painting?
Geodes are all the rage and with some resin, pigments, crushed stones, and glitter,
you too can make this glamourous style of art.

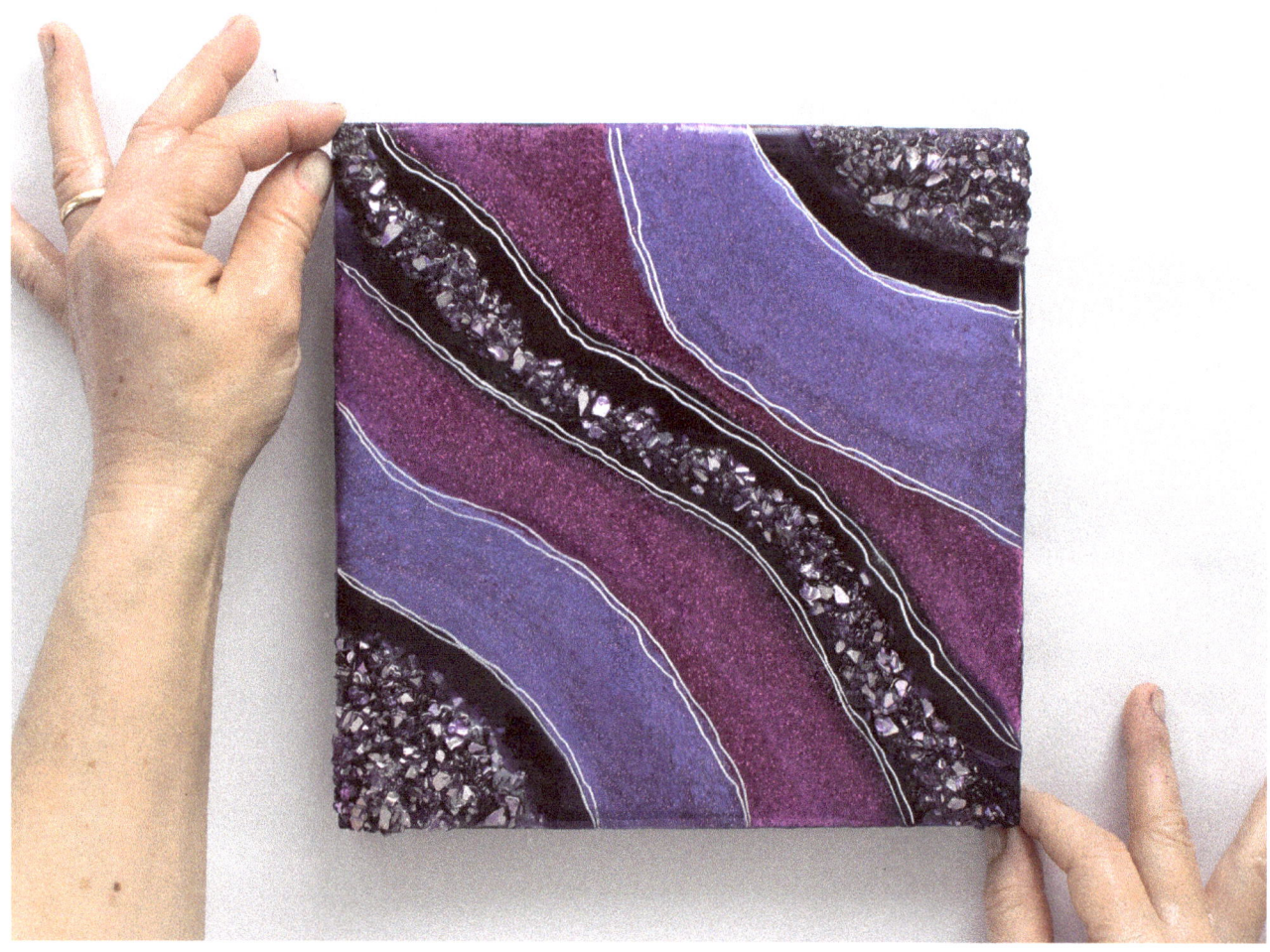

MATERIALS

sturdy substrate such as wood panel

pencil

painter's tape

black garbage bag or tablecloth

level

protective gloves

resin kit

measuring and mixing cups

resin colorant (tint, alcohol ink, mica powder)

craft sticks

glitter

crushed stones

butane torch

paint pens

box or container to cover your artwork

sandpaper or mini sander

acrylic paint for sides (optional)

1. Geode paintings can take several days to several weeks to create due to the curing time between the layers of resin and the size of the painting you make. Start with a smaller size panel until you feel comfortable creating something bigger. The key to creating a successful result is time, patience, and layers!

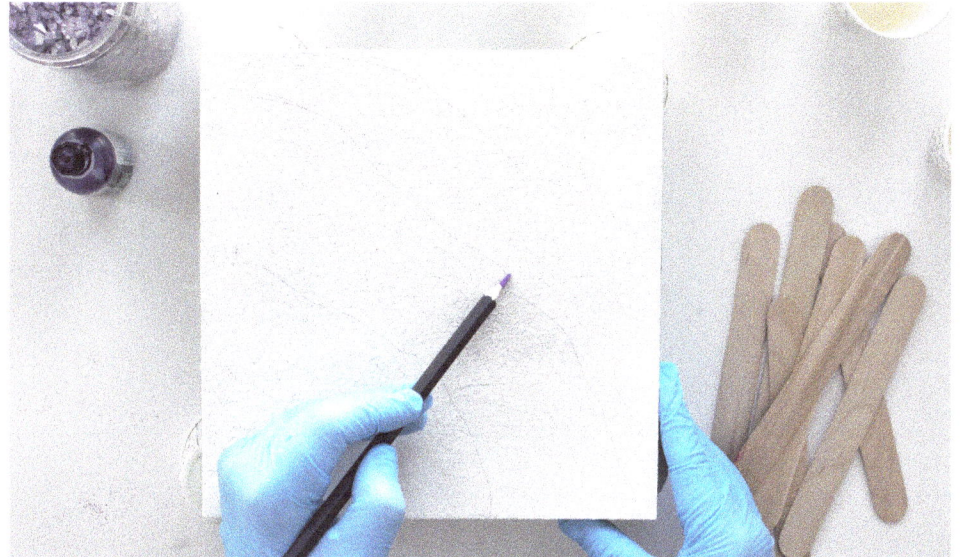

TIP

Use old artwork panels you created but didn't like. Simply paint over them with gesso and repurpose them.

2. With your pre-primed wood panel, mark out lightly with a pencil guidelines for your colors. Decide on which colors you're going to use before starting the project. You will be adding glitter to your tinted resin. Keep it simple with two or three colors. Apply tape to the sides of the substrate.

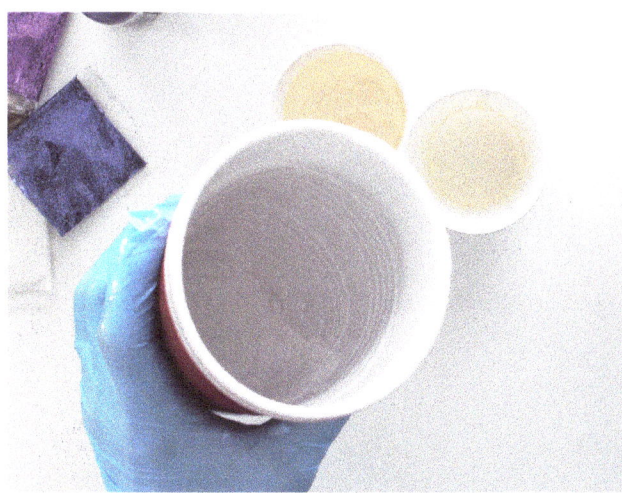

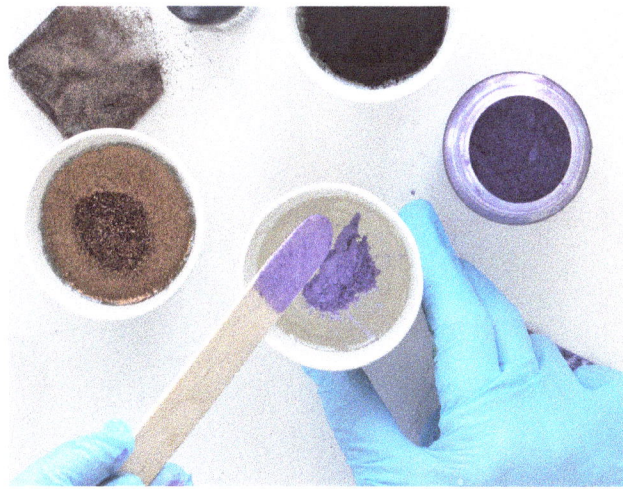

3. Work on a well-covered, level surface. Wearing protective gloves, mix up your resin according to the product instructions and pour into several containers. Mini paper bathroom cups work well for this project because you can bend the cups to form a spout for better control when pouring. You only need a few ounces (55 to 85 g) of resin for a 6" (15.2 cm) panel. Remember, you are working in layers!

4. Mix in your resin colorant with a craft stick. Use tints, alcohol ink drops, or mica powders. A drop or two is all that you need because these products are highly pigmented. Mix your resin well and add some glitter.

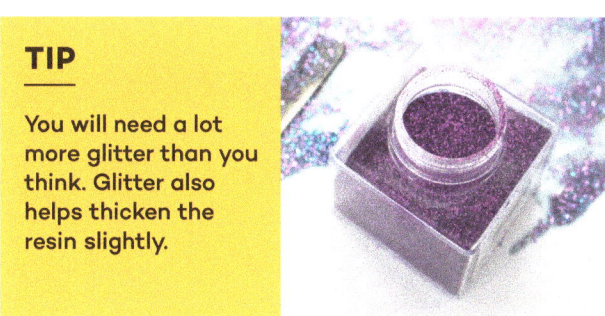

TIP

You will need a lot more glitter than you think. Glitter also helps thicken the resin slightly.

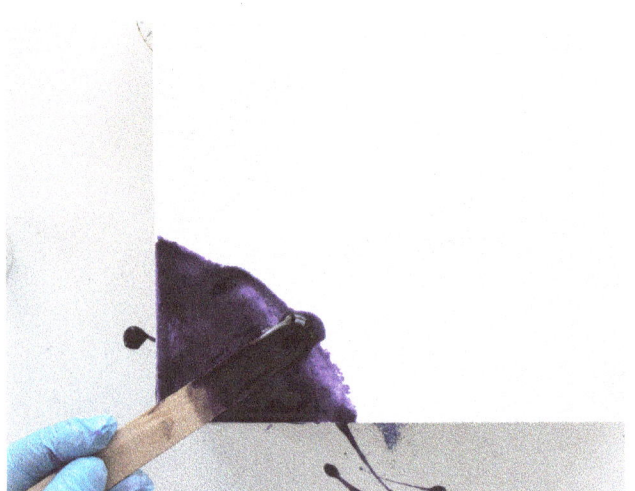
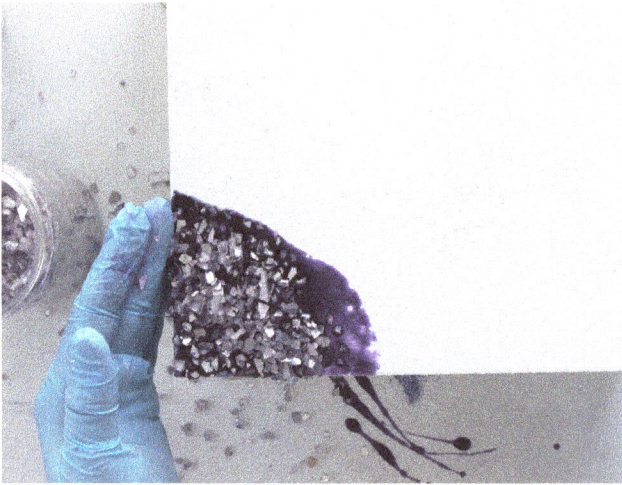

5. Begin by putting some colored resin onto the corner with your craft stick and spread over the marked area. Crushed stones look good on the corners, so carefully place some into the resin and move into place with your craft stick. Don't worry if all stones don't stick. You will be covering them with another layer of resin.

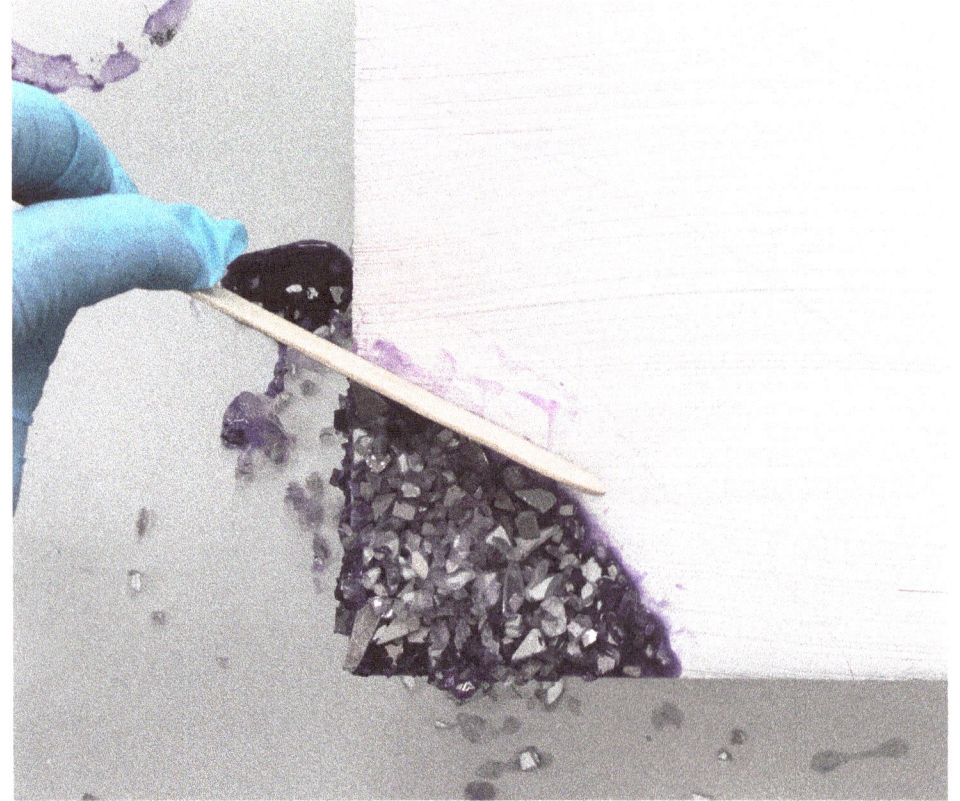

TIPS

- Make sure your panel is not sitting directly on your covered surface. Rest each corner of the panel on four upside-down mini paper cups. Use a level to ensure the panel is flat and even.

- Save money by purchasing clear stones. Color them yourself by adding the stones to a container and dropping alcohol ink into them. Mix and allow to dry.

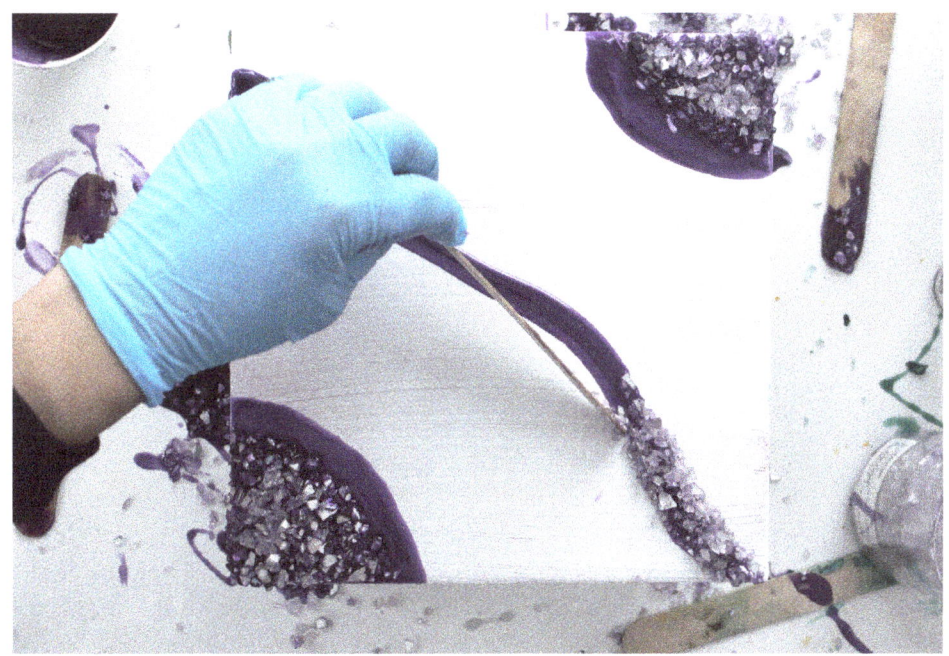

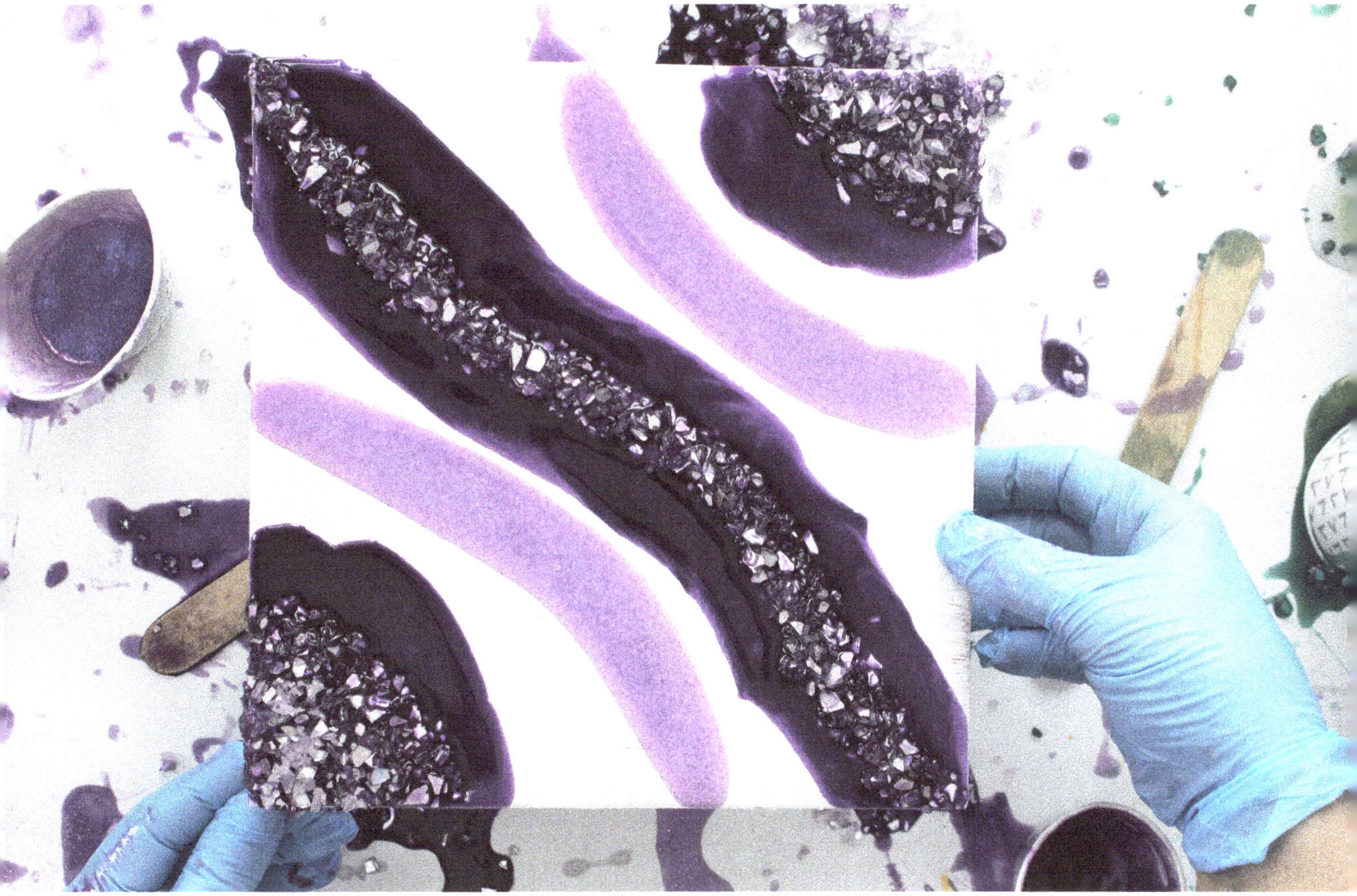

6. Continue to pour a center line across the panel and repeat with more stones. Next, pour another resin color across your pencil guide mark. Try to keep each colored pour minimal because the resin will spread and blend.

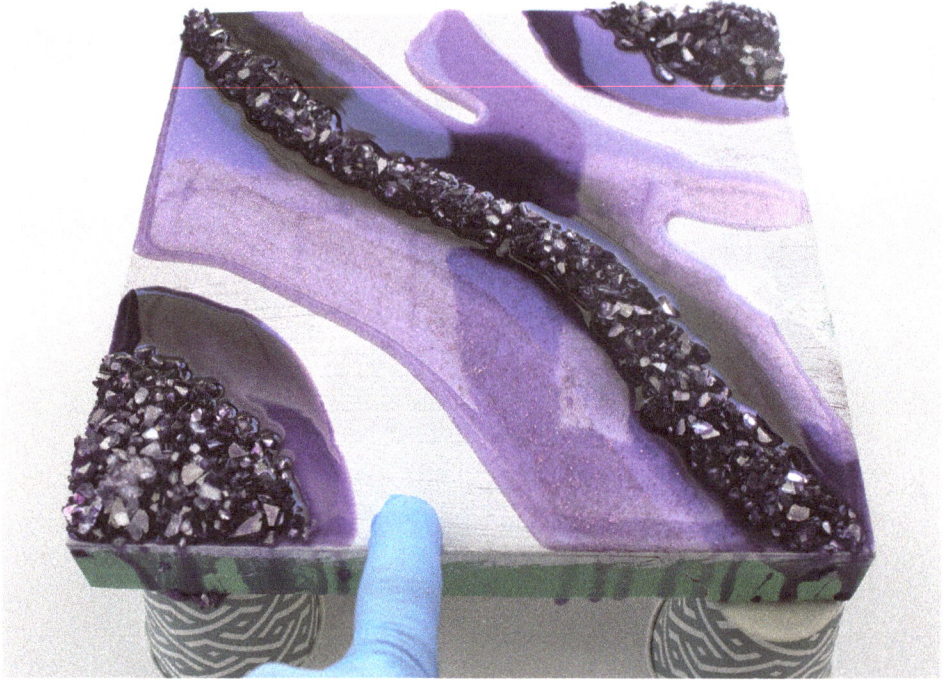

TIPS

• Allowing the resin to sit and thicken slightly will prevent the pour line from spreading too wide and too thinly.

• Applying multiple coats of resin can create problems when it comes time to remove the tape on your side panel. To make it easier, remove and reapply tape after each cured layer.

7. To create a complete separation of colors without blending, you *must* leave a bare section in between colors and allow the resin to semi-cure (3 to 5 hours) before pouring a new color next to it. This is more time-consuming but required if you don't want a blended effect. Don't worry if your first layers aren't perfect. You can fix them on the next layer.

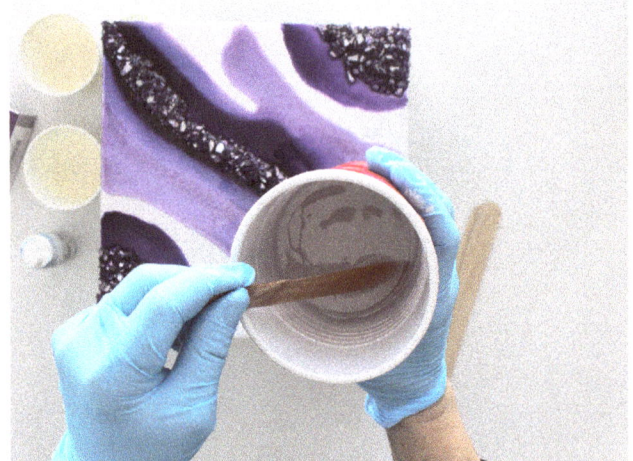
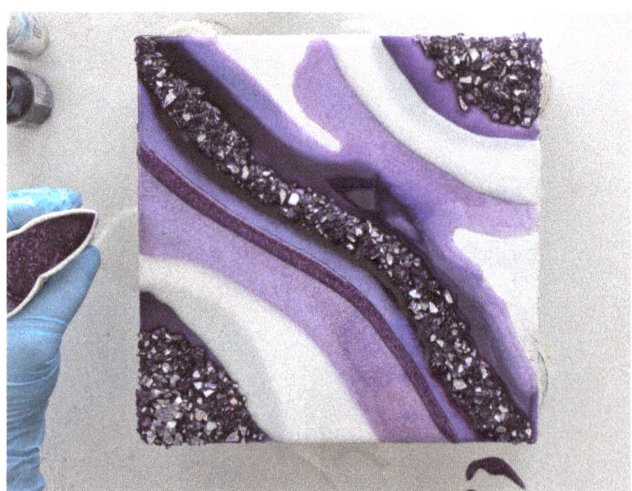

8. Once your layer has cured for the minimal amount of time for it to semi-harden, mix up another small amount of resin and begin the process over by filling in more layers until you are happy with the look. If you don't like a color, you can always change it by covering over with a new color. Layers can always be changed.

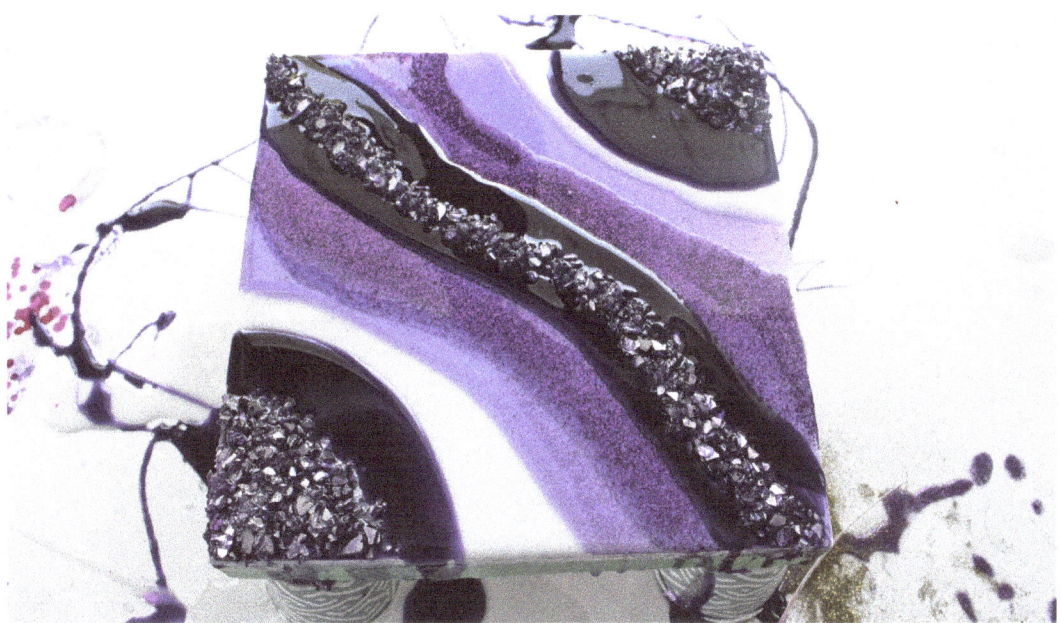

9. Torch if necessary (see project 2 for instructions on how to use the butane torch) and allow your panel to cure. Cover it overnight to prevent dust from settling in the resin.

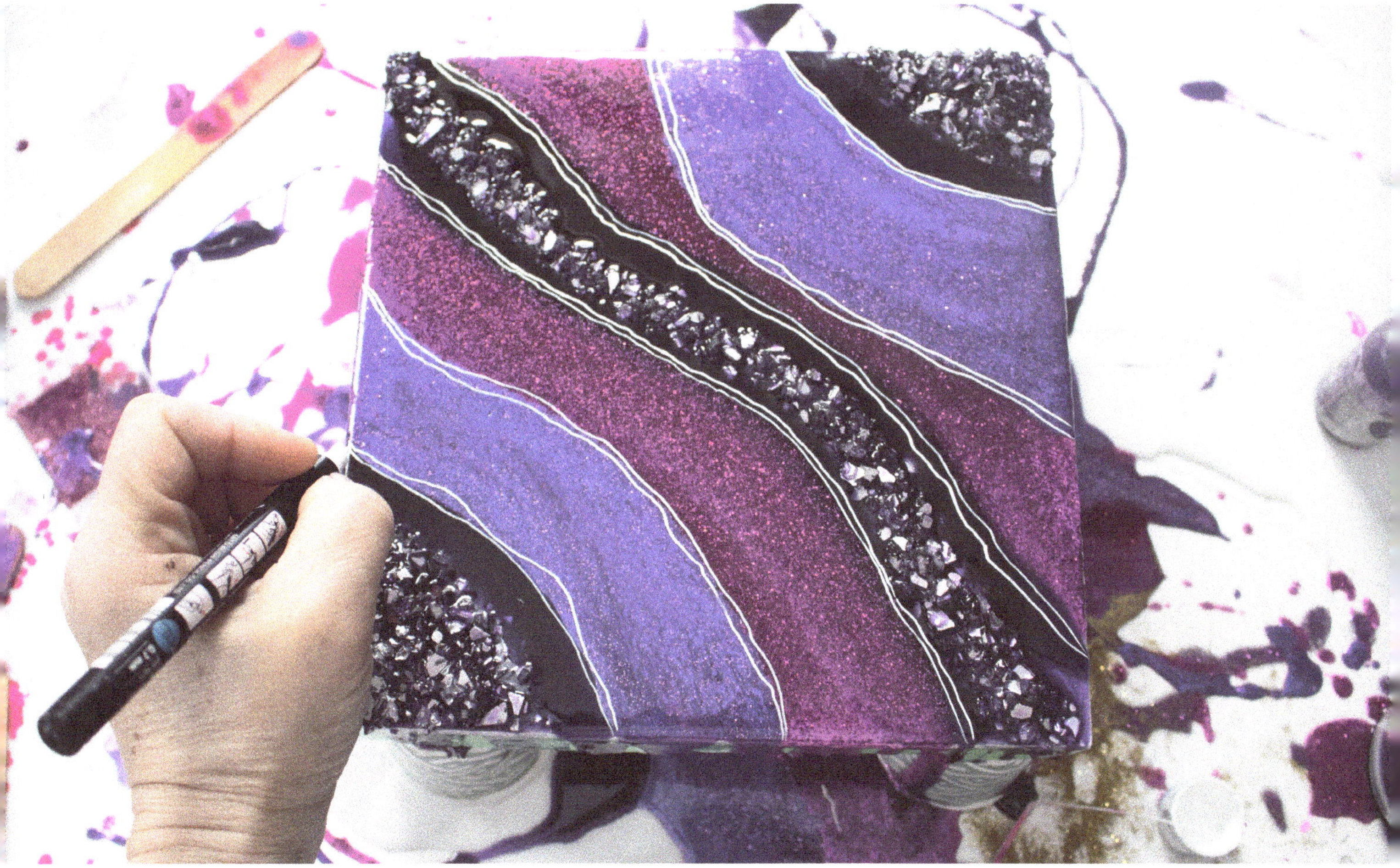

10. Once your panel has fully cured, add some fine line embellishing with a marker if you'd like.

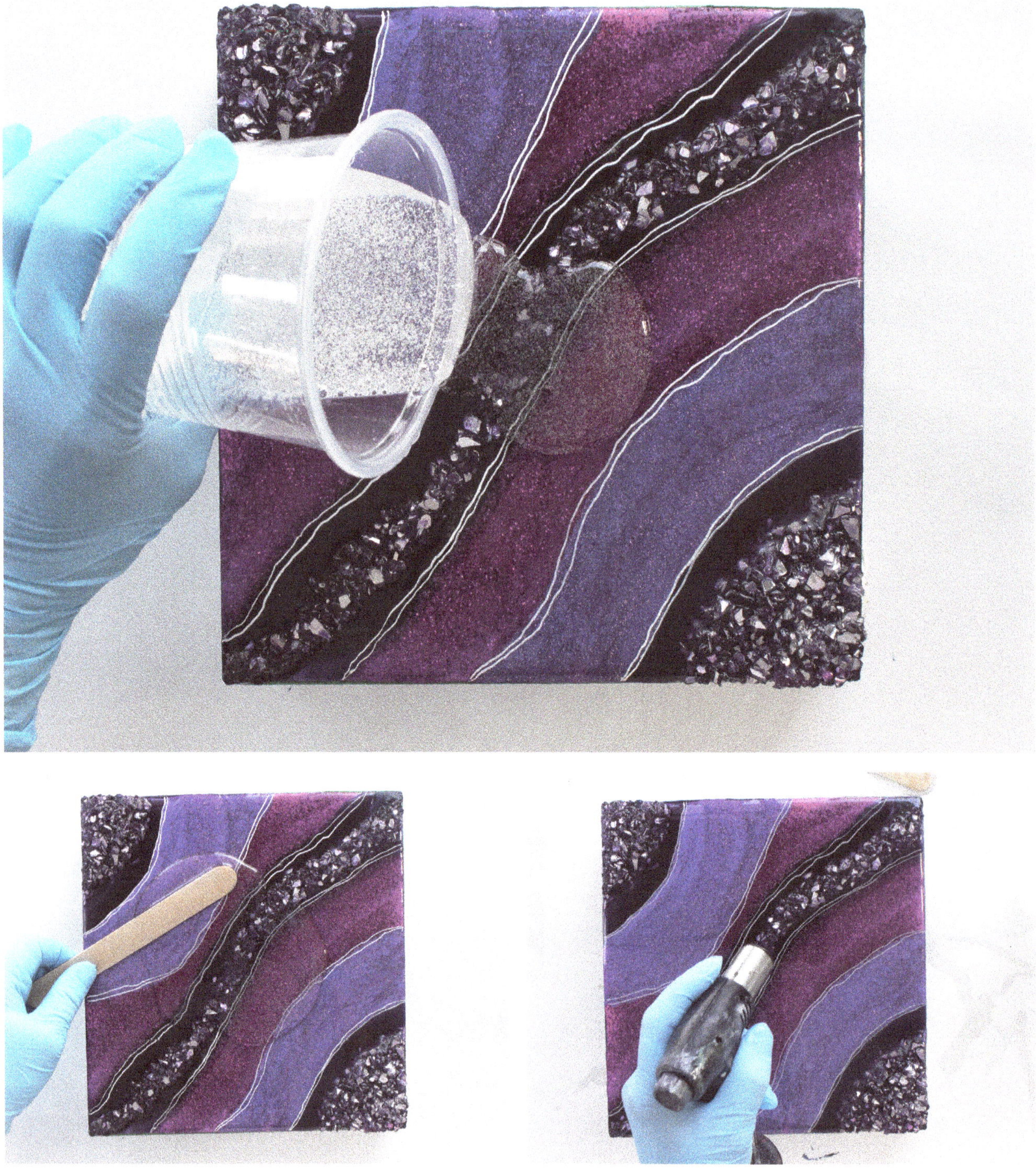

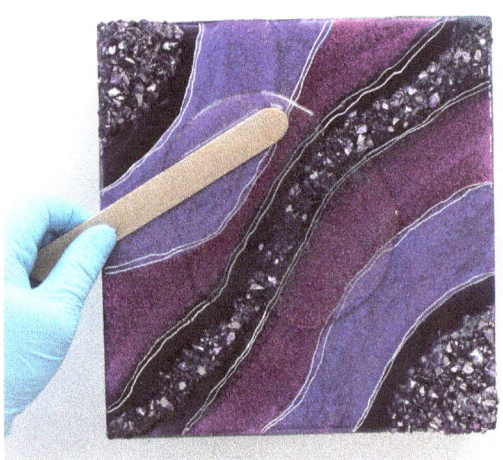

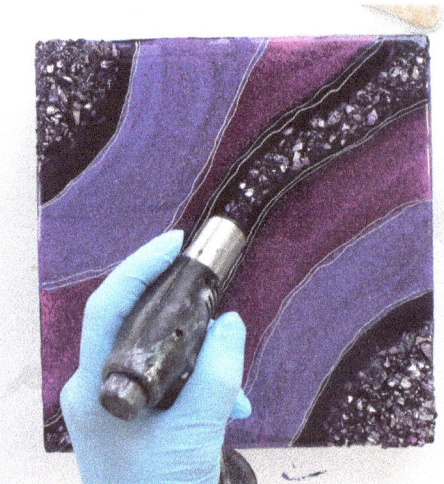

11. Measure and mix up a final batch of resin and pour over your entire painting. Torch to eliminate bubbles. Allow to cure, covered, for at least 12 to 15 hours before coming back to remove the side tape.

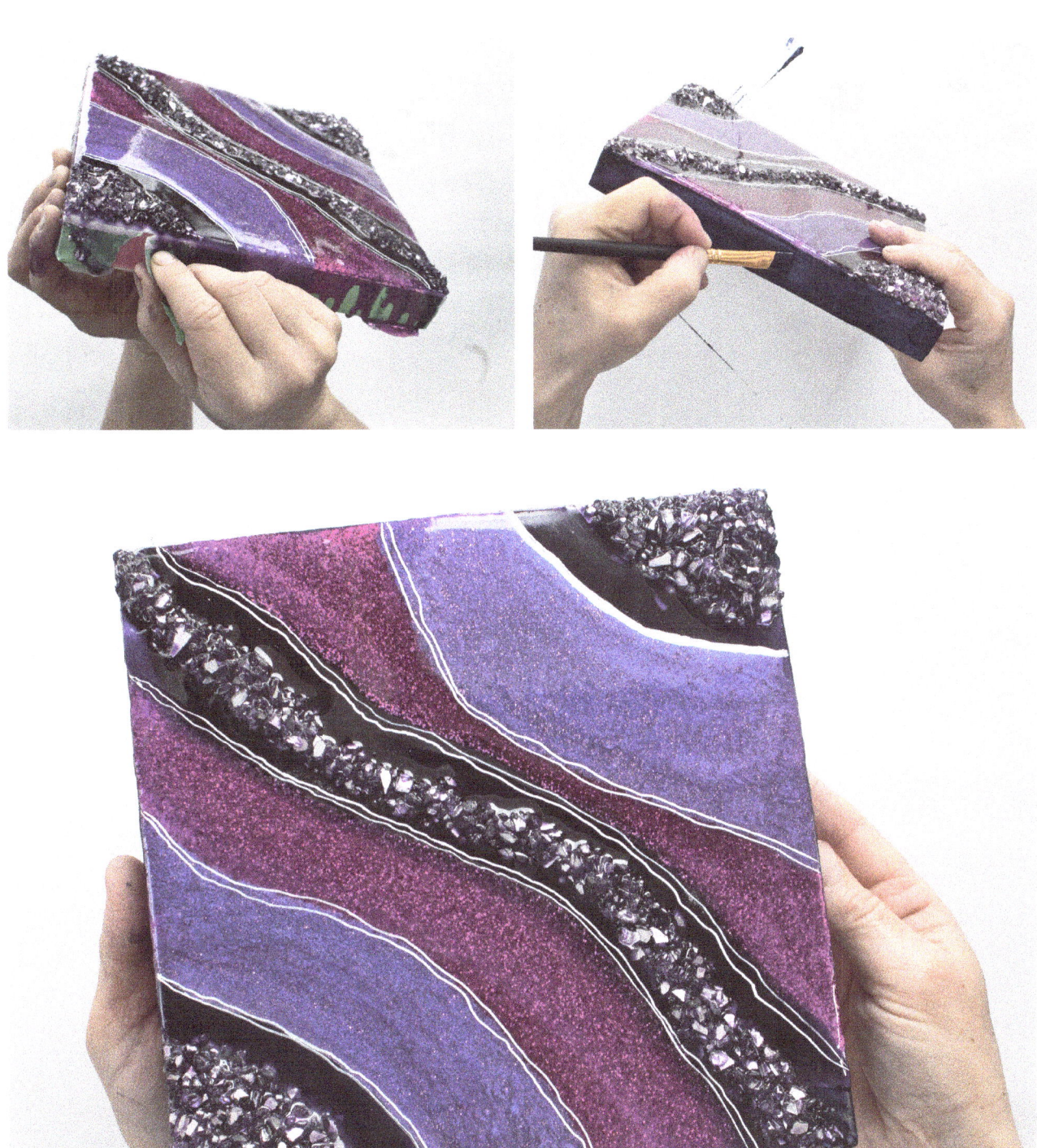

12. To finish your painting, remove the tape and sand by hand or use a mini electric sander for a smooth finish. Leave your sides natural or prime and paint with a coordinating acrylic paint color to match your geode.

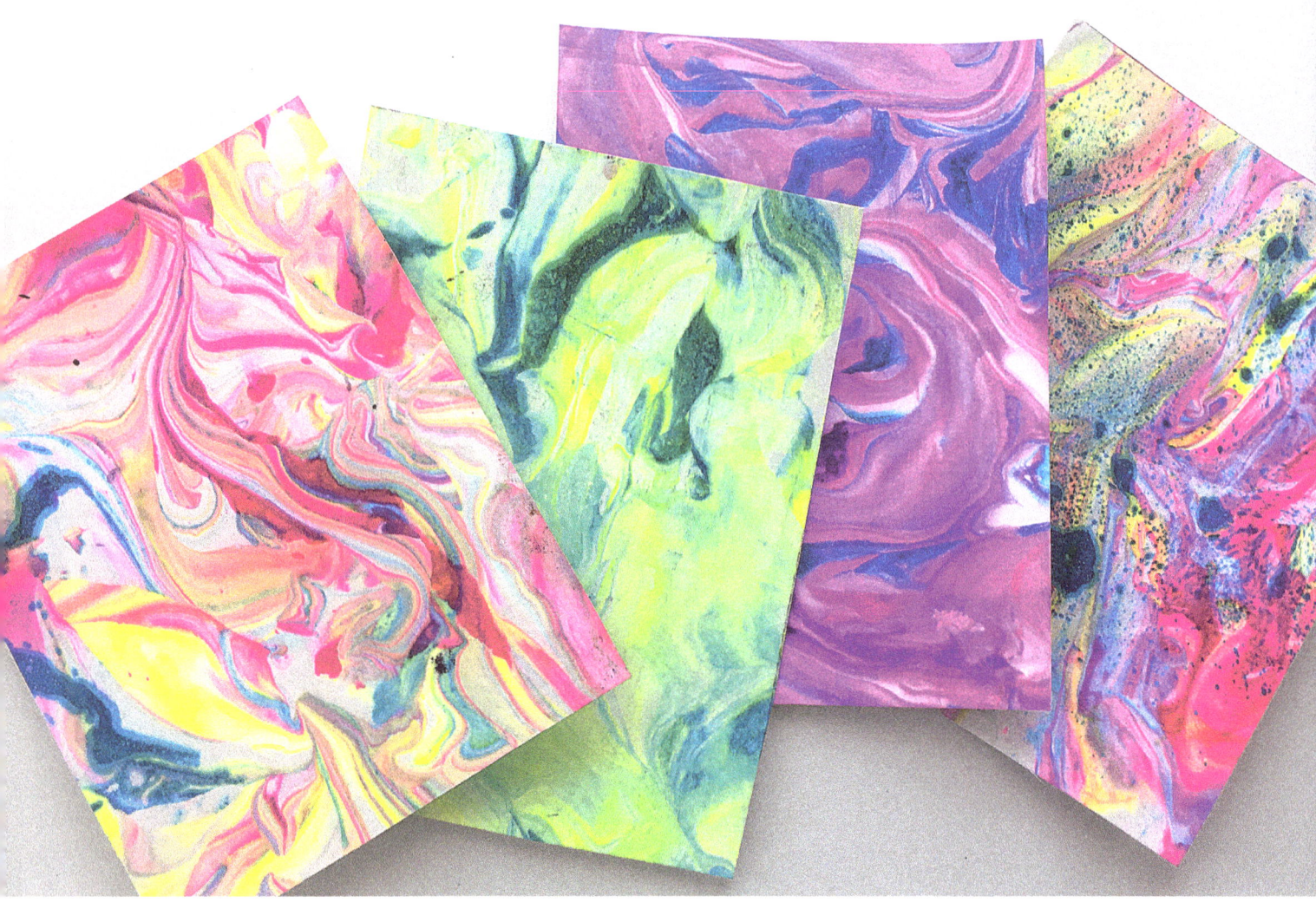

PROJECT 12

PAPER MARBLING WITH ALCOHOL INK

Marbling has been around for centuries and is still a popular form of art today. The colors and patterns can be bold and vibrant or soft and pretty. The oldest paper-marbling technique is called *suminagashi* (floating ink), which dates to the twelfth century in Japan. *Ebru*, originally from Turkey, is another form of intricate marbling in which detailed art is drawn with tools into a viscous solution and then transferred to an absorbent paper. Marbling properly on water requires several ingredients and a lot of preparation to the solution, but the technique shown here is easy, fun, and can be done in mere minutes!

MATERIALS

can of foam shaving cream (not gel)

large shallow container

scraper/spreading tool

alcohol inks

some thick paper (watercolor cardstock is ideal)

scissors

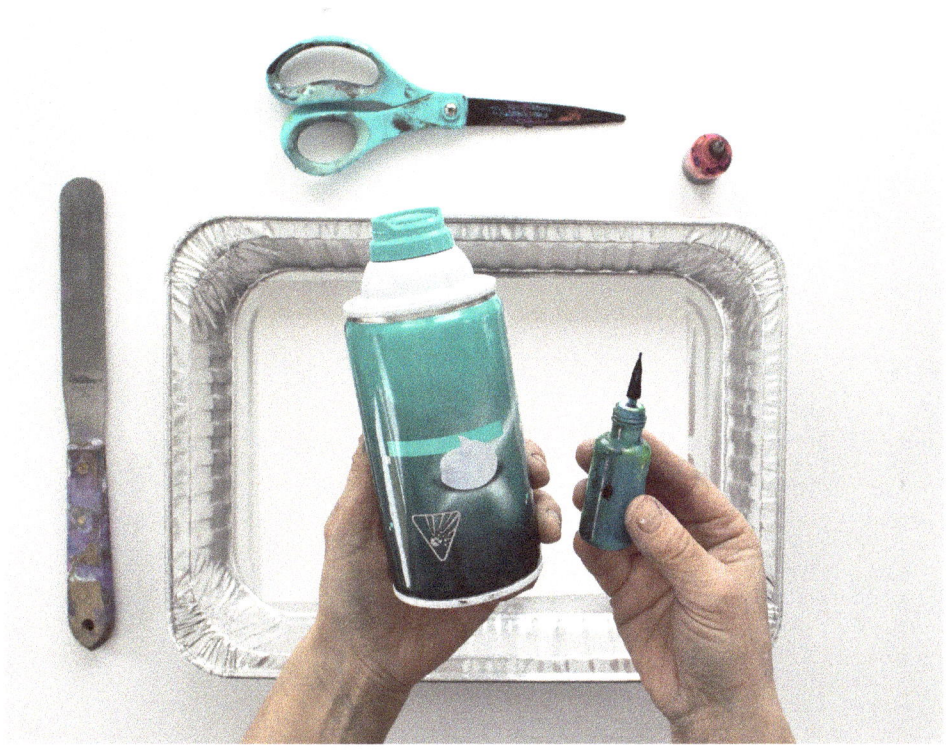

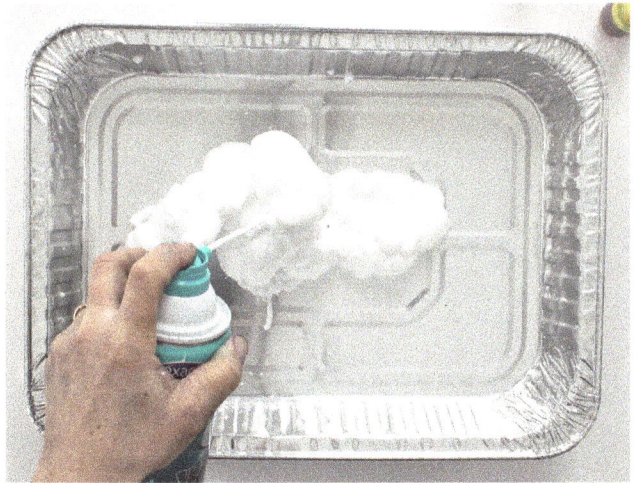

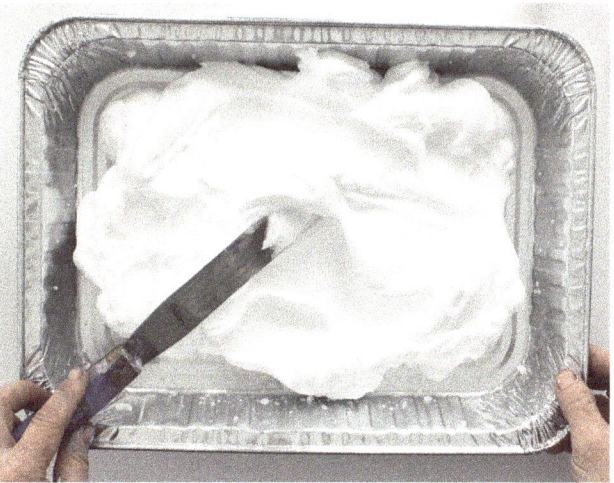

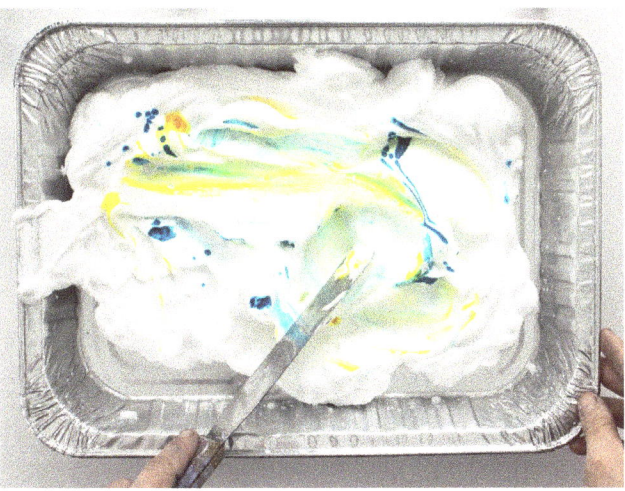

1. Begin by spraying some shaving cream into your shallow container. Spread it around a little with your scraper or spreading tool. Drop some alcohol ink colors onto the foam and lightly swirl it around until the colors intermingle.

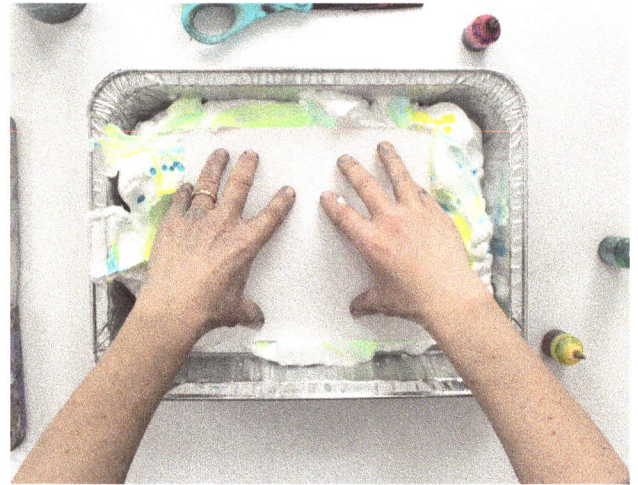

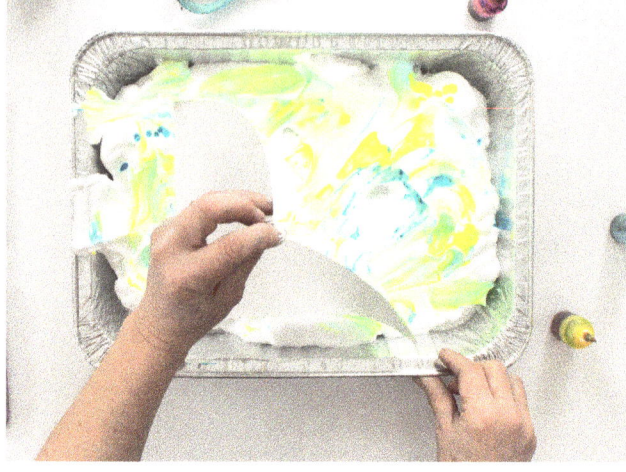

2. Gently press your paper down onto the shaving cream, making sure all areas are sitting on the mixture. Wait a few seconds before pulling up one corner and removing it from the surface. You will be left with a foam-covered paper.

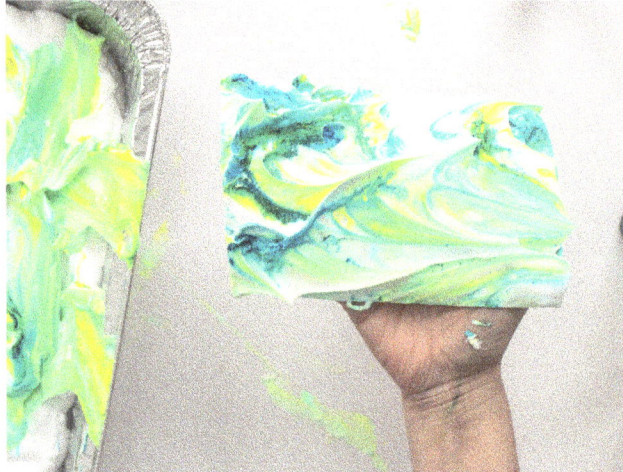

NOTE

Be careful not to scrape too hard or you will leave lines in your pattern.

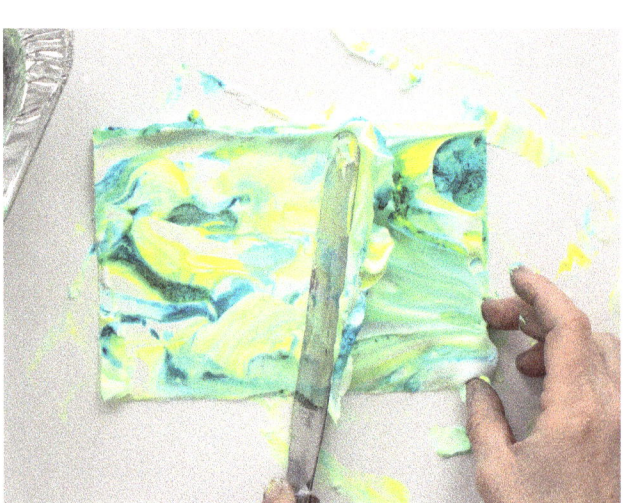

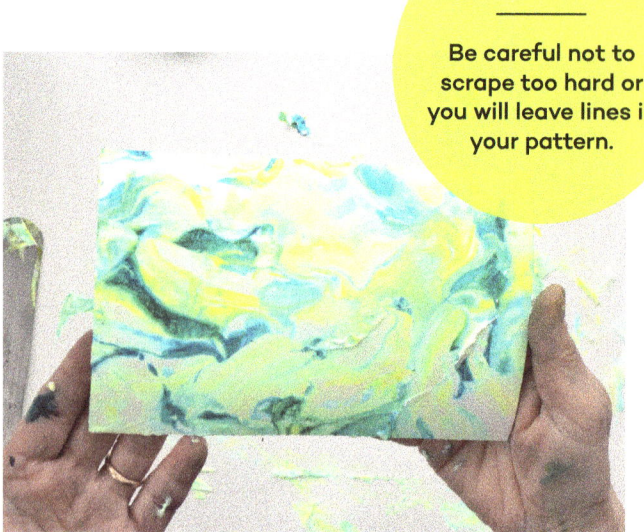

3. Place the cardstock down on a flat surface and with one hand holding the paper, lightly scrape off the shaving cream, revealing a marbled pattern transfer.

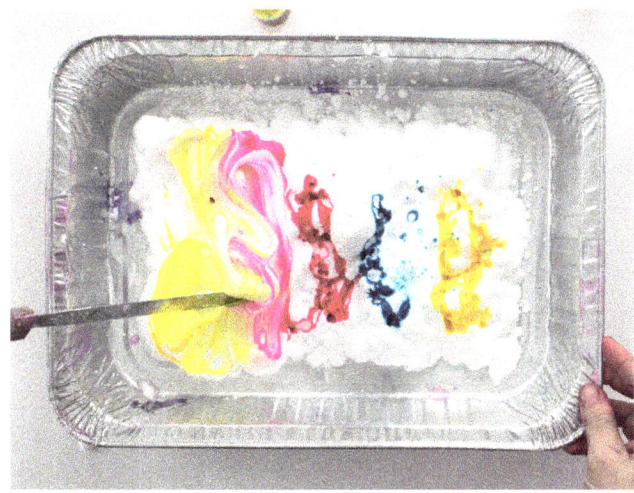

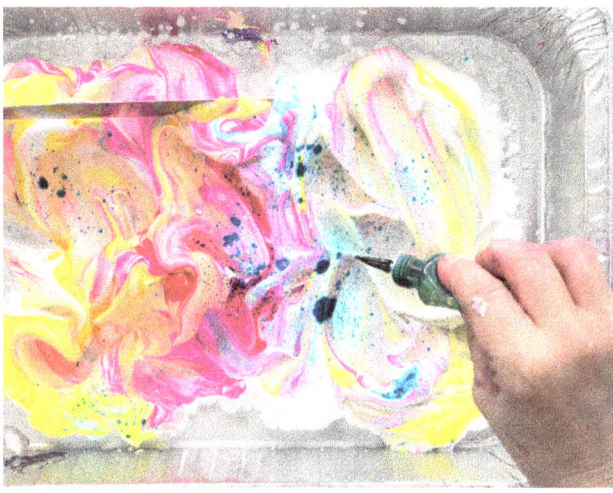

4. Repeat the process by either adding more colors to the existing foam tray or washing it out and beginning again with a new color combination. You can get creative and make a rainbow of colors by adding just blue, yellow, and pink! The more colors you add to the shaving cream, the more colorful the pattern.

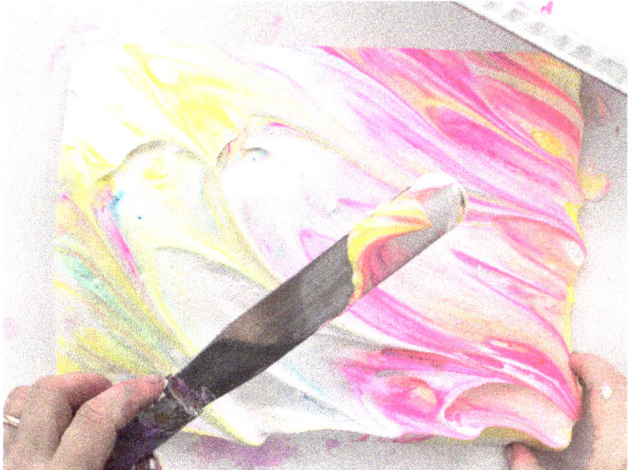

5. When all your colored patterns are created, allow them to dry for approximately 5 to 10 minutes. Then cut up the papers and use them for cardmaking, scrapbooking, and even embossing!

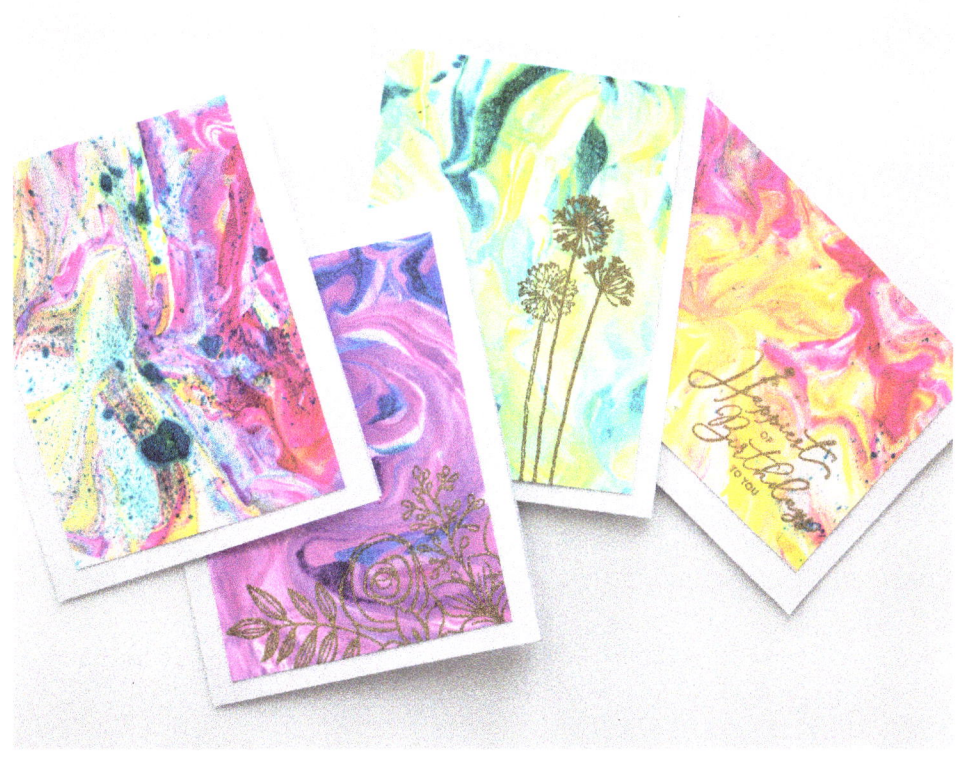

RESIN FAQ/ TROUBLESHOOTING

There are many brands and types of resin on the market, each offering its own benefits. The type purchased (polyester, epoxy, polyurethane) will determine the cure time (time it takes to become solid) and working time (mixing and pouring).

It is important to use a resin that is made specifically for your needs. All the projects in this book use an epoxy resin. You will also want one that has low odor and zero VOCs (volatile organic compounds). Many brands of resin on the market, such as polyurethane, are intended for industrial use (e.g., fiberglass boats), are toxic, have strong chemical smells, and yellow quickly. Be sure to do your research before you buy.

For your safety, always work in a well-ventilated area, wear gloves, and wear a certified approved mask against airborne organic vapors.

Resins are made up of a two-part epoxy system (one part base and one part hardener). Depending on the brand, you will mix and measure either by weight or by volume.

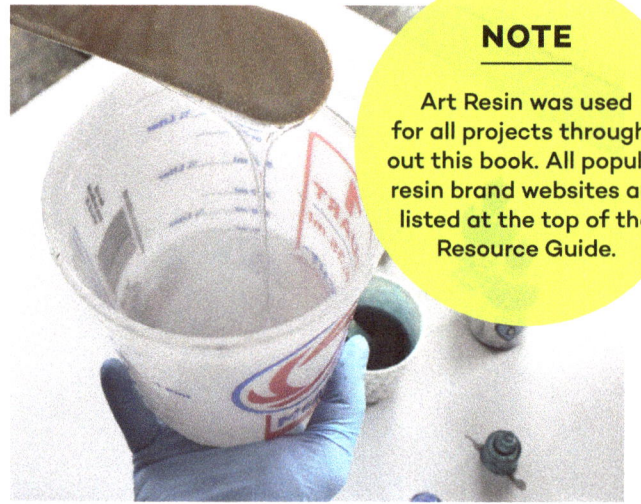

NOTE

Art Resin was used for all projects throughout this book. All popular resin brand websites are listed at the top of the Resource Guide.

There is a slight learning curve with resin, and everyone experiences this, no matter what type of resin is used. The most commonly asked questions, issues, remedies, and preventive steps are outlined here to help you problem solve.

ISSUE	CAUSES	REMEDY
Resin is not curing; still sticky/gooey	• Improper measuring/mixing • Too much colorant added (resin tint/alcohol ink)	• Read and follow instructions accurately. • Do not add more than 10 percent colorant/tint to resin mixture. • Scrape off uncured resin. Wipe surface and recoat.
Tiny bubbles cured in resin	• Poured too thick a layer • Didn't torch to release bubbles • Didn't seal artwork prior to resin • Mixed resin too quickly	• Do not pour more than ⅛" (3 mm) thick layers. • Use a butane torch to help eliminate bubbles. • Seal artwork first, especially raw wood. • Slowly mix resin.

ISSUE	CAUSES	REMEDY
Dents/dimples	• Fluctuating temperatures • Over-torching • Dust particles	• Work in a consistent temperature of 71°F to 75°F (22°C to 24°C). • Do not hold or torch repeatedly over the same area of the resin. Move quickly and evenly across the surface. • Cover artwork with a container while curing.
Bare spots/repelled from sides	• The surface area was oily	• Handle artwork with gloves to prevent natural oils from hands from transferring onto the sides and surface. • Sand with medium-grit paper and recoat.
Resin appears cloudy	• Water in resin	• Even a few drops of water that make it into resin can cloud it. Keep away from water.
Streaking/lines	• Manipulating past working time • Resin has begun to cure while still spreading	• Pour mixture out of container all at once. • Do not continue to scrape excess onto surface. • Leave resin alone when working time has been reached.
Oily cured surface	• Film caused by amines in hardener while curing in cooler temperatures	• Use a cloth and gently wash with warm water and dish soap.
Scratches/minor imperfections	• Accidental	• Wait 24 hours and sand surface with medium-grit sandpaper before recoating.
Drips on underside of panel/canvas	• Resin running down the sides while curing	• Carefully warm with a heat gun to ease cured drips off with a knife. • Tape and mask the underside of your artwork.
Resin cured on sides	• Resin pouring over edge	• Use an electric hand sander to remove hardened resin and for smooth finishes.
Settled dust particles	• Leaving resin to cure uncovered	• Use cardboard boxes or plastic containers to cover and protect during curing process.

- Don't be afraid to sand! Lightly sanding between cured layers gives the surface tooth and allows for good adhesion of the resin. Marks will not show when you reapply another coat of resin.

- Use silicone tools and containers when using resin. Leave a craft stick in an empty scraped-out container. Pulling out the stick the following day will bring with it the leftover cured resin film. You will be left with a clean, reusable container.

HEAT GUN VS. FLAME TORCH:

WHAT'S THE DIFFERENCE?

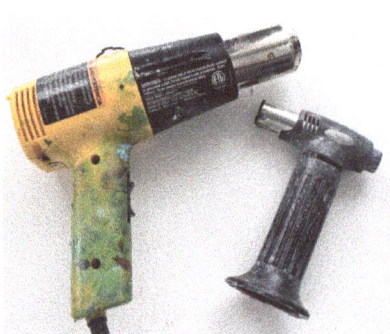

Both a heat gun and a torch (butane/propane) can be used with resin to eliminate bubbles. Which one you choose is determined by the results you desire. If you are using resin as a clear top coat finish on your art, you should opt for a torch. By gently moving the flame over the surface (barely kissing the resin), you will release the bubbles rising to the surface without moving or blowing around the resin itself.

If your desired result is to add depth, color blending, lacing, and texture to your resin, then you should use a heat gun. This tool blows hot air across the surface and literally pushes the resin around the surface, creating the opposite effect of a butane flame torch.

NOTES

1. Sometimes one bottle of your resin kit is yellow or yellows slightly once opened. This happens automatically over time and when the hardener is exposed to air. The chemical reaction causes it to yellow. This is normal and doesn't mean your resin is bad. However, you should avoid using resin that has been sitting unused for more than 12 months, as resin does have a shelf life.

2. All resins will yellow over time; however, today's science and technology have allowed brands to improve and delay this process by adding UV protection and HALS (hindered amine light stabilizers) to their products. HALS help interrupt the chemical interaction and block the yellowing process.

3. ALL artwork is prone to fading when hung in direct sunlight, regardless of which medium was used. Protect and hang your art away from harmful rays.

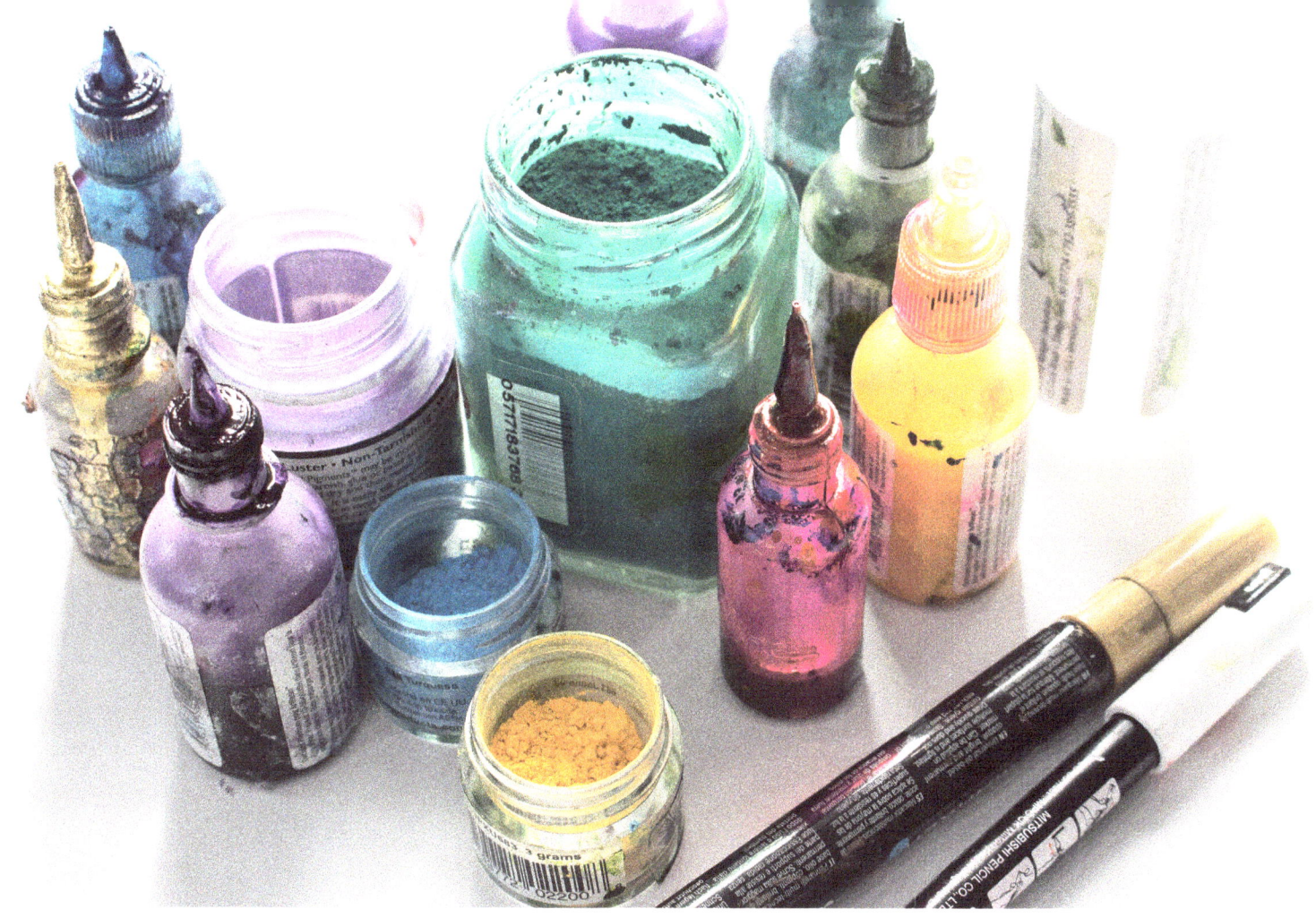

RESOURCE GUIDE

If you are unable to source products locally or online through your country's Amazon site, you can contact, shop, and search for retailers directly from the manufacturing brands listed below.

POPULAR ALCOHOL INK BRANDS

Jacquard
www.jacquardproducts.com

Ranger
www.timholtz.com

Brea Reese
www.breareese.com

Spectrum Noir
www.spectrumnoir.com

Copic
www.copic.jp

POPULAR RESIN BRANDS

Art Resin
www.artresin.com

Glasscast
www.easycomposites.co.uk

ResinPro
www.resinpro.it

StoneCoat
www.stonecoat
countertops.com

Barnes Epoxy
www.barnes.com.au

Just Resin
www.justresin.com.au

Crystal Clear
www.smooth-on.com

Artworks Resin
www.artworksresin.ca

Resin Obsession
www.resinobsession.com

EcoPoxy
www.ecopoxy.com

MasterCast
www.resinworks.ca

CounterCulture
www.counterculturediy.com

MICA POWDERS AND RESIN TINTS

Art Resin
www.artresin.com

Armour Art
www.armourart.com

Counter Culture
www.counterculturediy.com

Jacquard
www.jacquardproducts.com

Just Resin
www.justresin.com.au

Demco
www.demcoencouleurs.com

POPULAR ALCOHOL INK PAPERS (polypropylene)

Yupo & Yupo Translucent
www.yupo.com

TerraSlate
www.terraslatepaper.com

Dura-Lar
www.grafixarts.com

PROTECTIVE SEALERS, SPRAYS, AND ADHESIVES

(Note: Always test products before applying to finished work)

Krylon
www.krylon.com

Montana
www.montana-cans.com

Golden
www.goldenpaints.com

Sennelier
www.sennelier-colors.com

DecoArt/Americana
www.decoart.com

FolkArt/Mod Podge
www.plaidonline.com

Aleene's
www.aleenes.com

Liquitex
www.liquitex.com

3M
www.3m.com

ONLINE ARTIST WOOD PANEL AND CANVAS SUPPLIERS

Cass Art
www.cassart.co.uk (UK)

Curry's
www.currys.com (Canada)

Deserres
www.deserres.com (Canada)

Ampersand
www.ampersandart.com (Europe/Canada/U.S.)

Artist & Craftsman Supply
www.artistcraftsman.com (U.S.)

Rex Art
www.rexart.com (International)

POPULAR EMBELLISHING PAINT MARKERS AND PENS

Pebeo
www.pebeo.com

Krylon
www.krylon.com

Posca
www.posca.com

LAMPSHADE USED IN PROJECT ON PAGE 56:

8" (20 cm) diameter with 25" (63.5 cm) circumference white linen shade from www.bouclair.com
(Note: This project can be made with any alternative shade of your choice.)

POPULAR FLUID ART FACEBOOK GROUPS

Alcohol Ink Art Community

Alcohol Ink & Everything Else

Let It Flow

Resin Art Worldwide

Fluid Art & Paint Pouring

Acrylic Pouring Basics

Jane Monteith's online classes
www.janesclassroom.com

CLOSING REMARKS

Thanks for having fun with the projects in this book. I hope they inspired you to create some great things! When students take my online courses, I'm always amazed at what they create. I'd love to see what you created from this book. Sharing is half the fun, and you should be proud of everything you make. So, don't be shy: Say hello and post your pics on Instagram and tag me @janelovesdesign so I can see your beautiful projects.

Stay inspired and keep experimenting. You never know where it may take you.

xoxo,
Jane

ABOUT THE AUTHOR

Jane Monteith is a self-taught international-selling mixed-media artist and an online educator and influencer. While Jane was trained in advertising and marketing, she worked for more than a decade across Ontario as an independent artist designing and hand painting illustrations on vinyl substrates with highly pigmented screening inks. With her love of design and vibrant color combinations, she transitioned into working and creating her own art several years ago. Her work features colorful contemporary fluid patterns on birch wood panel, glazed with epoxy to result in a brilliant finish. Sharing her love of color, texture, and fluid art forms has gained Jane a large fan following on many social media platforms, including Instagram, Pinterest, and YouTube. Jane's popular MOD Minis (a term she coined when she started creating fluid ink collage paintings on 4-inch [10 cm] wood panels) have been sold worldwide and sell out monthly on her website. Jane's much larger original works are sold through KOYMAN Galleries, located in Ottawa, Ontario, Canada. She immigrated from England to Canada in 1981 and now lives just north of Toronto. For more on Jane, please visit:

www.instagram.com/janelovesdesign
www.janemonteith.com
www.youtube.com/janemonteith
www.janesclassroom.com

ACKNOWLEDGMENTS

I always dreamed of becoming a published author, so when Mary Ann Hall, editorial director of Quarry Books, approached me to write one, I was floored. My initial thought was maybe she got me confused with someone else and had made a mistake. There are thousands of fabulous artists and creative minds in this world, and I'm humbled and honored to be thought of as one of them. Thank you, Mary Ann, for believing in me and telling me I was THE artist. Thanks also to Marissa Giambrone, art manager and creative layout extraordinaire, who assured me several times that my photos looked great and produced a layout design beyond my expectations.

I wrote this book over the course of several snowy months seated on my couch by my front window alongside my dog Lucy. A few extra days were taken here and there at various ski hills during my son's freestyle competitions and again in miscellaneous airports, including Chicago, Aspen, Grand Junction, and Dallas, to name a few, while waiting patiently to arrive back home in Toronto.

I took hundreds of process photos for every single project in this book, which isn't always easy to do when working alone, especially with fluid art. Thank goodness for camera mounts and self timers! Although some professional photographers would probably shudder at the current state of my camera with its paint-stained body and sticky shutter release.

A high five and a big hug to my husband, Don, for turning a blind eye to my constant art mess and sometimes unrecognizable work studio, which is situated just off the kitchen in our home. Next dinner is on me. ;-)

To my loyal social media followers who support and love my work. My Instagram community is the foundation on which my online art journey began and I'm certain how this book opportunity came to be. Thank you for buying it and continuing to make me feel that what I do matters.

Lastly, a word about artists in general. Art brings people together and the world would be a boring place without it. Those artists who share their creativity and knowledge freely for all to see are indeed generous, inspiring, and uplifting. And for that, we should all be truly grateful.

Have fun and stay creative!
—Jane Monteith

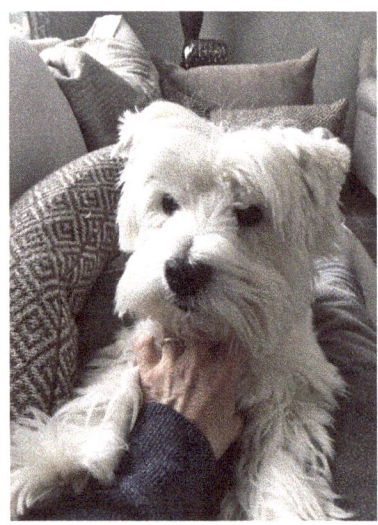
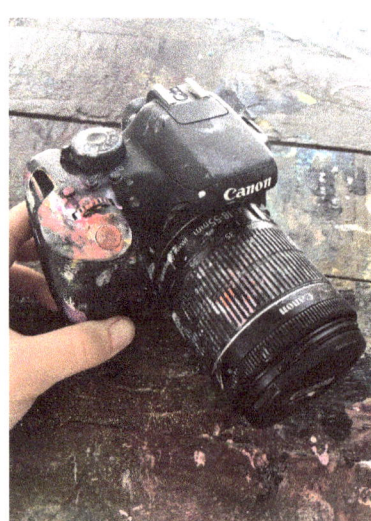
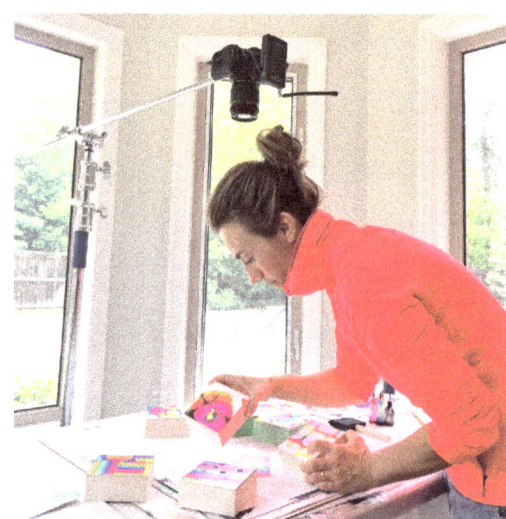

INDEX